GREEN PRODUCTIVITY IN SMALL AND MEDIUM ENTERPRISES

By the Same Author

Small Scale Industries (In 2 Volumes)

About the Editor

Dr. M. Soundarapandian is Professor of Rural Industries and Management and Director, Centre for Entrepreneurship Development, Gandhigram Rural Institute (Deemed University), Gandhigram. Apart from 22 years of teaching and extension responsibilities at Gandhigram Rural Institute, he has been engaged in reserch activities.

He has published many research articles in various journals at national and international level and eighteen books, namely, Rural Industries in India, Women Entrepreneurship, Tribal Development, Industrial Resurgence in Rural India, Literacy Campaign in India, Rural Entrepreneurship, Research Projects in Rural Industries and Management, Rural Industries Under Liberalisation, Rural Industrialisation, Small Scale Industries, Economic Reforms & Public Distribution System, Management Model for Self-employment, Economic Reforms and Employment, Environment Degradation and Rural Poverty, Textile Industry under Globalisation, Rural Labour Market etc.

Seven scholars were awarded Ph.D. Degree under his guidance and supervision. Apart from providing research guidance to a number of Ph.D. and M.Phil Scholars, he has directed and completed eleven minor Research Projects and four major Research Projects that were sponsored by University Grants Commission, Indian Council of Social Science Research, Bharat Gyan Vijayan Samiti, Ministry of Rural Development, New Delhi and State Resource Centre for Non-formal Education, Chennai. The Young Economist Award by the Association of Economists of Tamil Nadu and Bharat Jothi Award by the India International Friendship Society have been awarded to him.

GREEN PRODUCTIVITY IN SMALL AND MEDIUM ENTERPRISES

Volume II

(Environment, Health and Waste Management)

Edited by

M. Soundarapandian
Professor of Rural Industries and Management,
Director,
Centre for Entrepreneurship Development (CED)
Gandhigram Rural Institute
(Deemed University)
Gandhigram–624302

CONCEPT PUBLISHING COMPANY, NEW DELHI-110059

Cataloging in Publication Data—DK
Courtesy: D.K. Agencies (P) Ltd. <docinfo@dkagencies.com>

Green productivity in small and medium enterprises / edited by M. Soundarapandian.
2 v. cm.
Papers presented at a national seminar held at Gandhigram; with special reference to India.
Includes bibliographical references.
Includes indexes.
Contents: v. 1. Industry and agriculture — v. 2. Environment, health and waste management.
ISBN-13: 9788180693700 (set) ISBN-10: 8180693708 (set)
ISBN-13: 9788180693717 (v. 1) ISBN-10: 8180693716 (v. 1)
ISBN-13: 9788180693724 (v. 2) ISBN-10: 8180693724 (v. 2)

1. Small business—Environmental aspects—India—Congresses. 2. Agriculture—Environmental aspects—India—Congresses. 3. Environmental health—India—Congresses. 4. Refuse and refuse disposal—India—Congresses. I. Soundarapandian, M. (Mookkiah), 1960-

DDC 338.642 22

ISBN-13 : 978-81-8069-370-0 (Set) ISBN-10 : 81-8069-370-8 (Set)
ISBN-13 : 978-81-8069-371-7 (Vol-I) ISBN-10 : 81-8069-371-6 (Vol-I)
ISBN-13 : 978-81-8069-372-4 (Vol-II) ISBN-10 : 81-8069-372-4(Vol-II)

First Published 2007

Published and Printed by
Ashok Kumar Mittal
Concept Publishing Company
A/15-16, Commercial Block, Mohan Garden
New Delhi-110059 (India)

Phones : 25351460, 25351794
Fax : +91-11-25357103
Email : publishing@conceptpub.com

Dedicated

to

My Mother

Samuthiram

PREFACE

Green Productivity is not a new concept but has emerged gradually. Because it concentrates upon the preventive approach, it assumes the utmost importance as far as small-scale industries are concerned. Green productivity can be explained as adoption and implementation of environmentally sound processes and procedures to optimize the utilization of resources enhance production. In other words, this concept focuses upon enhancing producing without rating adverse impacts on the environment. Seen from a different perspective, improvement of the environment should have beneficial rather than delerious effects on the quantity and quality of production. Thus green productivity stresses optimum utilization of space, manpower, energy and raw materials, i.e., all the factors of production.

In view of strict implementation of existing environment regulations and more regulations to be promulgated in the future, two things have become clear. First, it will be difficult for small and medium enterprises to continue to operate. For example, in India many small-scale units have been forced to close or change location for environmental reasons. Second, if SMEs invest in waste (end of pipe) treatment, their production costs may not remain competitive. Now SMEs are finding them-selves, in a dilemma, and the wayout appears to be the green productivity approach because it result in a clearer environ-ment and higher production at less cost.

Green productivity can resolve a number of major problems of SMEs pertaining to the environment. It can enable them to sustain today's cut-throat competition and survive against multinational companies. On the other hand, it will help industries to create a positive image as an environmentally responsible company in the market, while at the same time helping the company to conform to regulatory requirements. Once the concept of green productivity is understood, the question arises of how to attain green productivity and make it a perpetual movement.

Based on the above background, a national seminar was organized with the assistance of AICTE, New Delhi.

Selected papers presented in the seminar are edited as a book on "Green Productivity in Small and Medium Enterprises". The book has been divided into four parts, viz., Industry, Agriculture, Waste Management and Environment and Health. The book is very useful to the young researchers, academicians and the policy-makers.

I am thankful to Dr. T. Karanakaran, Vice-Chancellor and Dr. N. Meenakshisundaram, Professor and Head, Department of Rural Industries and Management for their encouragement in completing the work. It is my pleasure to thank my wife Mrs. S. Indumathi and my children S. Diwakaran and S.Vidya for their kind cooperation. Finally, I am grateful to the Concept Publishing Company, New Delhi for publishing the book.

M. Soundarapandian

CONTENTS

LIST OF TABLES & FIGURES

Figures

LIST OF THE CONTRIBUTORS

Dr. A. G. Murugesan
Division of Industrial Toxicology and Pollution Control
Sri Paramakalyani Centre for Environmental Sciences
Manonmaniam Sundaranar University, Alwarkurichi-627412.

A. Inigo Antony Beaula
Research Scholar
Department of Rural Industries and Management
Gandhigram Rural Institute
Gandhigram-642302

A. Padrakali
S.G. Lecturer
A.P.C.M. College
Tuticorin.

B. Nirmala Devi
Research Scholar, Department of Management Studies
Manonmaniam Sundaranar University.
Tirunelveli.

Dr. B. Rajasekaran
H.O.D. Department of Management Studies
Manonmaniam Sundaranar University.
Tiruneveli.

Dr. C. Madhavi
Faculty member in Business Administration
Annamalai University
Annamalai Nagar-608002.

Dr. C. Samudhara Rajakumar
Faculty member in Business Administration
DDE, Annamalai University
Annamalai Nagar-608002.

C. Sunitha
Lecturer (S.S.)
Department of Economics
K.N.G.A. College, Thanjavur.

Dr. D. Ayub Khan Dawood
Department of Commerce
Quaide Milleth College
Medavakkam
Chennai - 601 302.

G. Narasimha Raghvan
Research Scholar,
Department of Economics,
PSG College of Arts and Science,
Coimbatore -14.

G. Sathis Kumar
M. Phil Scholar
Department of Economics
Madras Christian College
Chennai-600059.

Dr. I. Sundar
Lecturer in Economics
Directorate of Distance Education
Aannamalai University
Annamalai Nagar.

J. Kavitha
Lecturer, Department of Management
Sankara College of Science and Commerce
Coimbatore-641 035.

J. Rajesh Khanna
Lecturer,
Department of Management Studies and Commerce,
Bharath College of Science and Management, Thanjavur-5.

K. Kavitha
Division of Industrial Toxicology and Pollution Control
Sri Paramkalyani Centre for Environmental Sciences
Manonmaniam Sundaranar University, Alwarkurichi-627412.

K. Tamizh Jyothi
Faculty member in Business Administration
DDE, Annamalai University
Annamalai Nagar-608002.

M. Karthikeyan
Project Fellow
Department of Cooperation
Gandhigram Rural Institute
Gandhigram.

M. Rajadurai
Head, Department of Environmental Economics
School of Economics
Madurai Kamaraj University
Madurai - 625 021.

Dr. M. Sivasubramanian
Reader in Business Administration
Directorate of Distance Education
Annamalai University
Annamalai Nagar

Dr. M.S.K.N. Nagendran
H.O.D. of Commerce
Sri Kaliswari College
Sivakasi.

Dr. M. Soundarapandian
Professor,
Department of Rural Industries and Management
Gandhigram Rural Institute
Gandhigram-624302.

Dr. N. Malathi
Professor,
Department of Economics,
Annamalai University
Annamalai Nagar.

Dr. N. Meenakshi Sundaram
Head, Department of Rural Industries and Management
Gandhigram Rural Institute,
Gandhigram.

P. Ayyammal
Research Scholar
Department of Economics
Gandhigram Rural Institute
Gandhigram-624302.

P. Jagadeesh
Project Staff
Department of Cooperation
Gandhigram Rural Institute
Gandhigram.

P. Natarajan
Principal,
Coimbatore Institute of Management and Technology
Narasipuram, Thondamuthur
Coimbatore - 641 109.

P. Selvamani
Faculty of Rural Social Science
Gandhigram Rural Institute,
Gandhigram.

R. Meena Devi
Research Scholar
Department of Environmental Economics
Madurai Kamraj University
Madurai-21.

R. Renganayaki
Research Scholar
Department of Rural Industries and Management
Gandhigram Rural Institute
Gandhigram.

S.B. Bhosale
M. Phil Research Scholar,
Department of Economics
Shivaji University
Kolhapur-416004.

Dr. S. Dhandapani
Head, Department of Mechanical Engineering
Coimbatore - 641 014.

S. Kannan
Reader, Kamraj College,
Tuticorin.

Samwel Kakuko Lopoyetum
Post Doctoral Research Fellow
Department of Cooperation
Gandhigram Rural Institute
Gandhigram.

Dr. S. Padmavathi
Reader
Department of Economics
K.N.G.A. College, Thanjavur

Dr. S. Ramaswamy
Reader, Department of Economics
Gandhigram Rural Institute
Gandhigram-624302.

Dr. S.R. Choodambigai
Reader in Economics
Avinashilingam University
Coimbatore-43.

S. Sivakumar
Research Scholar
Department of Political Science and Development Administration
Gandhigram Rural Institute
Gandhigram.

S. Srinivasa Raghvan
SSN School of Management & Computer Applications
SSN Nagar, Old Mahabalipuram Road, Kalavakkam-603110.

S.V. Saravanan
Assistant Professor
Department of Mechanical Engineering
Sri Ramakrishna Engineering College, Coimbatore - 641 002

Srinivasa Gumparthi
Assistant Professor
SSN School of Management & Computer Applications
SSN Nagar, Old Mahabalipuram Road, Kalavakkam-603 110.

S. Sathya Bama
Research Scholar in Economics
Annamalai University
Annamalai Nagar.

T. Saravanan
Research Scholar
Department of Rural Industries and Management
Gandhigram Rural Institute
Gandhigram - 624302.

V. Dhulasi Birundha
Professor
Department of Environmental Economics
Madurai Kamaraj University
Madurai–21.

Dr. V. Mohanasundram
Reader, Department of Economics
PSG College of Arts and Science
Coimbatore-14.

V. Pugazhenthi
Sr. Lecturer & Head
Department of Commerce
Govt. Arts College, Ariyalur.

V.S. Karthik
Research Scholar
Department of Politial Science and Development Administration
Gandhigram Rural Institute
Gandhigram.

V. Sudhakar
Lecturer in Commerce
Sri Kalisvari Collage
Sivakasi

Dr. V. V. Mahamuni
Reader, Department of Economics
Modern College, Vaishnavi
Mumbai.

1

DISPOSAL OF SOLID WASTE IN KOLHAPUR CITY

A Suggestive Approach

V.V. Mahamuni and *S.B. Bhosale*

In this paper we will study the characteristics of sample households, opinions of the sample households regarding solid waste management (SWM) made in Kolhapur Municipal Corporation (KMC) and their suggestions for proper waste disposal.

OPINIONS OF SAMPLE HOUSEHOLDS ABOUT SWM IN KOLHAPUR

The Opinions on Domestic Solid Waste Generation

An attempt has been made to quantify the solid waste generation in the KMC area. In order to know the state of solid waste in Kolhapur city, the survey was conducted. The Kolhapur Municipal area was divided into 5 different localities as per the housing development, roads and other infrastructure like telephone, water supply, electricity and markets etc. These five localities were as under :

(i) The High standrad of living localities like Nagala Park and Tarabai Park.
(ii) The new extended are as like Shaupuri and Rajarampuri.
(iii) The Old Kolhapur like Mangalwar Peth and Shivaji Peth.
(iv) The low standard of living area like Shahu Mill's Workers Colony and Sidharthnagar.
(v) Slum areas like Sadar Bazar and Rajendranagar.

From each locality, 25 households were selected as samples for the survey work. While doing so, random stratified sampling method was adopted. The heads of the sample households were asked how much solid waste was generated in their houses per day? The response is shown in Talbe 1.1. In all 125 sample households were surveyed and out of these 42 (33.6%) heads of the households said that they were generating in the range of 500 gms to 1000 gms, 32 (25.6%) heads of the sample households said that they were generating in the range of 1000 gms to 2000 gms, and 10 (8.0%) heads of the household told that they were generating

Table 1.1 : Generation of the Solid Waste as per the Locality in Kolhapur City

Sr. No.	*Name of the Locality*	*Upto 500 gm*	*501 to 1000 gm*	*1001 to 2000*	*2001 to 3000 gm*	*More than 3000 gm*	*Total*
1.	The high standard of living locality like Park and Tarabai Park (having maximum number of bungalows and RCC type of housing	7 (28.0)	6 (24.0)	9 (36.0)	2 (8.0)	1 (4.0)	25 (100)
2.	The new extended are a like Shahupuri and Rajarampuri (having maximum number of RCC and flat type of housing units)	12 (48.0)	8 (32.0)	4 (16.0)	-	1 (4.0)	25 (100)
3.	The old Kolhapur like Managalwar Peth and Shivaji Peth (having maximum number of old and Manglori roof type of housing units).	6 (24.0)	6 (24.0)	8 (32.0)	3 (12.0)	2 (8.0)	25 (100)
4.	The low standard of living area like Shahu Mill Workers Colony and Sidharthnagar (having maximum number of kachha and Manglori roof type of housing units).	9 (36.0)	7 (28.0)	5 (20.0)	4 (16.0)	-	25 (100)
5.	The slum area like Sadar Bazar and Rajendranagar (having maximum number of kachha and hut type of housing units)	8 (32.0)	10 (40.0)	6 (24.0)	1 (4.0)	-	25 (100)
	Total	**42 (33.6)**	**37 (29.6)**	**32 (25.6)**	**10 (8.0)**	**04 (3.2)**	**125 (100)**

Source : Field Survey

solid waste in the range of 2001 to 3000 grams, while 04 (3.2%) heads of the households replied that they were generating the solid waste more than 3000 grams per day.

The table also shows that there is no, any direct relationship between the locality and the volume of solid waste generated by the people, but on an average every household had generated the solid waste more than 500 gms per day.

Segregation of Solid Waste

The segregation of solid waste is very important step in the waste management. It was asked to the heads of the sample households whether they were segregating the solid waste as dry and wet and compostable and non-compostable. The responses are shown in Table 1.2. Out of 125 sample households, 44 (35.2%) heads of the sample household, told 'yes' and 81 (64.8%) replied as 'No', in

Table 1.2 : Segregation of Solid Waste as Dry and Wet and Compostable and Non-compostable

S. No.	*Name of the Locality*	*Dry and Wet*		*Compostable and Non-compostable*	
		Yes	*No*	*Yes*	*No*
1.	The high standard of living locality like Nagala Park and Tarabai Park (having maximum number of bungalows and RCC type of housing units)	6 (24.0)	19 (76.0)	6 (24.0)	19 (76.0)
2.	The new extended area like Shahupuri and Rajarampuri (having maximum number of RCC and flat type of housing units)	10 (40.0)	15 (60.0)	7 (28.0)	18 (76.0)
3.	The old Kolhapur like Mangalwar Peth and Shivaji Peth (having maximum number of old and Manglori roof type of housing units)	2 (8.0)	23 (92.0)	2 (8.0)	23 (92.0)
4.	The low standard of living area like Shahu Mill Workers Colony and Sidharthnagar (having maximum number of *kachha* and Manglori roof type of housing units)	1 (4.0)	24 (96.0)	2 (8.0)	23 (92.0)
5.	The slum area like Sadar Bazar and Rajendranagar (having maximum number of *kachha* and hut type of housing units)	25 (100.0)	00	25 (100.0)	00
	Total	44 (35.2)	81 (64.8)	42 (33.6)	83 (66.4)

Source: Field Survey.

case of dry and wet segregation. In case of segregation as compostable and non-compostable, out of 125 sample households 42 (33.6%) heads of the households said 'yes' and 83 (66.4%) said 'No'. The table shows that maximum people were not segregating the solid waste, and they threw waste into municipal refuse bin.

Storage of Solid Waste at Home

The storage is important component of solid waste management. In order to know how the storage was made in the house, the question was asked to the sample households. The storage of waste means plastic bins, bucket, tin pot and other things like boxes etc., were stored at house level. The distribution of waste by the sample households is shown in Table 1.3. Out of 125 sample households, 62(49.6%) heads of the sample households were using plastic bin 25(20.0%) sample households replied that they were using buckets, 19 (15. 2%) said that they were using tin-pots, and 19 (15.2%)

Table 1.3 : Storage Facilities of Solid Waste at Home

Sr. No.	*Name of the Locality*	*Plastic bins*	*Bucket*	*Tin pot*	*Other things like boxes*	*Total*
1.	The high standard of living locality like Nagala Park and Tarabai Park (having maximum number of bungalows and RCC type of housing units)	16 (64.0)	9 (36.0)	-	-	25 (100)
2.	The new extended are like Shahupuri and Rajarampuri (having maximum number of RCC and flat type of housing units)	13 (52.0)	5 (20.0)	3 (12.0)	4 (16.0)	25 (100)
3.	The old Kolhapur like Mangalwar Peth and Shivaji Peth (having maximum number of old and Manglori roof type of housing units,	21 (84.0)	3 (12.0)	-	1 (4.0)	25 (100)
4.	The low standard of living area like Shahu Mill Workers Colony and Sidharthnagar (having maximum number of *kachha* and Manglori roof type of housing units)	11 (44.0)	3 (12.0)	9 (36.0)	2 (8.0)	25 (100)
5.	The slum area like Sadar Bazar and Rajendranagar (having maximum number of *kachha* and hut type of housing units)	1 (4.0)	5 (20.0)	7 (28.0)	12 (48.0)	25 (100)
	Total	**62 (49.6)**	**25 (20.0)**	**19 (15.2)**	**19 (15.2)**	**125 (100)**

Source: Field Survey.

heads of the households said that they were using cardboard-boxes or plastic bags. So far as income level is concerned, high income group people use plastic bins and low income group use the buckets and boxes etc. This shows that every house has one or the other storage facility of their own.

Disposal of Plastic

The plastic material is found in large quantity in solid waste. The different types of plastics are very important component of solid waste which affects the environment. The plastic carry bags (more than 20 Microns) are also used on large scale. In the survey it was to the heads of households in which manner they disposed off their plastic waste. The responses are shown in Table 1.4. Out of 125 sample households 86(68.8%) heads replied that they threw it into the domestic waste, 26(20.8%) heads of the households told that they were selling it after segregation, 8(6.4%) heads of the sample household said that they threw plastic waste to rag pickers and hardly 5(4.0%) heads of the households replied that they burnt the plastic bags along with other waste. It was found that in the income category of Rs. 15000 no family was using plastic for reselling.

Lifting of Solid Waste by KMC

In order to know the state of collection (lifting) of solid waste it was asked to the sample househlds what was the frequency of refuse lifting trucks. The response of the respondents is shown in Table 1.5. Out of 125 samples, 53 samples heads of households (45.4%) told that everyday in the morning refuse collector trucks were collecting the solid waste, 27 households (21.6%) said that on alternate days refuse lifting was done, 36 households (28.8%) replied that once in a week, and 5 households (4.0%) told that once in fifteen days refuse lifting vehicles were visiting their area. It was found that some important localities and commercial areas had regular collection of solid waste; while the low income group localities were neglected by the KMC. It seems that the household

Table 1.4 : Disposal of Plastic

S.No. Name of the Locality	Throwing in waste	Selling after segregation	Giving to rag pickers	Burning	Total
1. The high standard of living locality like Nagala Park and Tarabai Park (having maximum number of bungalows and RCC type of housing units)	15 (60.0)	7 (28.0)	3 (12.0)	-	25 (100)
2. The new extended area like Shahupuri and Rajarampuri (having maximum number of RCC and flat type of housing units)	13 (52.0)	6 (24.0)	3 (12.0)	3 (12.0)	25 (100)
3. The Old Kolhapur like Mangalwar Peth and Shivaji Peth (having maximum number of old and Manglori roof type of housing units)	15 (60.0)	8 (32.0)	2 (8.0)	-	25 (100)
4. The low standard of living area like Shahu Mill Workers Colony and Sidharthnagar (having maximum number of *kachha* and Manglori roof type of housing units)	18 (32.0)	5 (20.0)	-	2 (8.0)	25 (100)
5. The slum area like Sadar Bazar and Rajendranagar (having maximum number of *kachha* and hut type of housing units)	25 (100)	-	-	-	25 (100)
Total	**86 (68.8)**	**26 (20.8)**	**8 (6.4)**	**5 (4.0)**	**125 (100)**

Source: Field Survey.

who might be the tax payers, were served in a better way by the Sanitary Department.

Adequacy of Refuse Bins

For proper solid waste management, the adequacy of refuse bins is very essential. In the survey it was enquired whether refuse bins were adequate in number or not. Out of 125 sample households, 39 (31.2%) respondents replied that refuse bins were adequate in number, while 86 (68.8%) heads of the sample households told that refuse bins were not adequate in number. It means that there was shortage of refuse bins in Kolhapur. The citizens are very

Table 1.5 : Lifting of Solid Waste by KMC

S.No	*Name of the Locality*	*Time of lifting solid waste*	*Every-day*	*Alternate days*	*Once in a week*	*Once in fort-night*	*Total*
1.	The high standard of living locality like Nagala park and Tarabai Park (having maximum number of bungalows and RCC type of housing units)	In the morning	9 (36.0)	8 (32.0)	7 (28.0)	1 (4.0)	25 (100)
2.	The new extended area like Shahupuri and Rajarampuri (having maximum number of RCC and flat type of housing units)	In the morning	9 (36.0)	5 (20.0)	4 (16.0)	4 (16.0)	25 (100)
3.	The Old Kolhapur like Mangalwar Peth and Shivaji Peth (having maximum number of old and Manglori roof type of housing units)	In the morning	18 (72.0)	4 (16.0)	3 (12.0)	-	25 (100)
4.	The old standard of living area like Shahu Mill Workers Colony and Sidharthnagar (having maximum number of *kachha* and Manglori roof type of housing)	In the morning	15 (60.0)	8 (32.0)	1 (4.0)		25 (100)
5.	The slum area like Sadar Bazar and Rajendranagar (having maximum number of *kachha* and hut type of housing units)	In the morning	2 (8.0)	2 (8.0)	21 (84.0)		25 (100)
		Total	**53 (42.4)**	**27 (21.6)**	**36 (28.8)**	**05 (4.0)**	**125 (100)**

Source: Field Survey.

particular about their inconvenience though they are not bothering to pay their annual house property taxes.

Adverse Effects of Solid Waste on Citizens

In order to know the effects of solid waste, it was asked heads of the sample households whether they were facing any bad effects of solid waste. The distributions of responses of the respondents were shown in Table 1.6.

Out of 125 sample households, 107(85.6%) heads of the households replied 'Yes' and 18(14.4%) heads replied 'No'. It means that majority of the people especially in lower income category were facing the evil effects of solid waste in Kolhapur. Out of 125 heads of the households 60(48.0%) heads of sample households said that they were facing the problem of odour and 28(22.4%) heads of the sample households told thay that were facing the problem of nuisance of the rats, pigs and animals, while 19(15.2%) heads said that they were facing all the above problems.

As regard to the diseases the question was put to the respondents but they could not exactly identify that the diseases were spread due to poor disposal and mismanagement of solid waste. But any citizen while passing by such bins can experience the bad odour by his nose and eyes.

The Role of Sanitary Inspector

The role of sanitary inspectors is crucial in solid waste management. In order to know the public opinion regarding the role of sanitary inspectors of KMC, three questions were asked, whether they know the name of Sanitary Inspectors of their areas, what were their duties and whether they were working honestly or not.

Out of 125 heads of the sample households, 96(76.8%) heads replied that they were not knowing the name of the Sanitary Inspectors. Out of 125 heads of sample households 85(68.0%) heads told that they did not know the duties of Sanitary Inspectors.

Regarding the honest working of Sanitary Inspectors, 23(92.0%) heads of the sample households said that the Sanitary Inspectors were working honestly. Thus there is a mixed reaction regarding the working of Sanitary Inspector. Now the Inspectors have provided the mobile phone facility for quick services.

Table 1.6 : Adverse effects of solid waste on citizens of Kolhapur

S.No.	*Name of the Locality*	*Effect*		*Type of Effect*				*Total*
		Yes	*No*	*Odour problem*	*Diseases*	*Nuisance of rats Pig, and animals*	*All types of Prob-lems*	*Yes+No*
1.	The high standard of living locality like Nagala Park and Tarabai Park (having maximum number of bungalows and RCC type of housing units)	19 (76.0)	6 (24.0)	13 (52.0)	-	2 (8.0)	4 (16.0)	19+6 =25 (100)
2.	The new extended areas like and Shahupuri Rajarampuri (having maximum number of RCC and flat type of housing units)	19 (76.0)	6 (24.0)	11 (44.0)	-	2 (8.0)	6 (24.0)	19+ 6=25 (100)
3.	The old Kolhapur like Mangalwar Peth and Shivaji Peth (having maximum number of old and Managlori roof type of housing units)	19 (76.0)	6 (24.0)	17 (68.0)	-	2 (8.0)	-	19+6 =25 (100)
4.	The low standard of living area like Shahu Mill Workers Colony and Sidharthnagar (having maximum number of *kachha* and Manglori roof type of housing units)	25 (100)	00 (00)	7 (28.0)	-	14 (56.0)	4 (16.0)	25 (100)
5.	The slum area like Sadar Bazar and Rajendranagar (having maximum number of *kachha* and hut type of housing units)	25 (100)	00 (00)	12 (48.0)	-	08 (32.0)	05 (20.0)	25 (100)
	Total	107 (85.6)	18 (14.4)	60 (48.0)	-	28 (22.4)	19 (15.2)	125 (100)

Source : Field Survey.

Nature of Co-operation to KMC from Sample Respondents

In order to know the participation (involvement) of the citizens in solid waste management, it was asked to the heads of sample households that how do they help to KMC in solid waste management. Out of 125 sample households, 105(84.0%) heads of the households said that by depositing solid waste into the municipal refuse bins, they were helping to KMC, and 5(4.0%) heads of the sample households said that they were helping by segregating the solid waste, while 4(3.2%) heads said that they were making the awareness among the neighbour families, and 11(8.8%) households replied that they were not paying attention to the solid waste management. It was noticed that higher income categories were more aware about the clean environment as they read newspapers and listen to media.

Cleanliness, Campaign and Recycling Efforts made by People and Organisations

In the proper solid waste management; Youth Clubs, Housing Society, Community Based Organisations (CBOs) and Non-Government Organisation (NGOs) can play crucial role. In order to know the role of these organisation, the question was asked to heads of the sample households whether there was any private cleanliness campaign in their locality. Out of total 125 sample households, 23(18.4%) heads said 'Yes' and 77(81.6%) heads of the sample households said 'No'. It means that in some localities efforts were going on; while in majority of localities efforts were not made. Overall picture of the data shows that there were no any private efforts of cleanliness and recycling of waste in Kolhapur. But the NGOs like Nisarg Mitra, Forum of Citizens for Save Kolhapur, WWF, Blind Belief Relief Committee, etc. are aware of the problems and they have taken some actions also in this regard by organising the programmes in schools and high schools.

Compost from Organic Waste

In the sample survey, it was also asked whether the citizen make compost from solid waste. Out of 125 heads of the sample households, only 15(12.0%) heads of the households said that they were making the compost from organic solid waste and use it in agriculture. But 110(88.0%) heads of the households said that they were not making the compost from waste. It was observed that in Kolhapur there was no recycling of solid waste except few agriculture families in Old Kolhapur.

Role of Pollution Control Board

In order to throw light on the role of Pollution Control Board, the question was asked whether the District Pollution Control Board has taken the action against the huge solid waste generation those who were not managing it properly in the Kolhapur city or not. Out of 125 heads of the sample households, 121(96.8%) heads said that they had not heard that Pollution Control Board had taken any action against any citizen or any otheer organisation who has thrown aways the solid waste other than authorised places. This shows that Pollution Control Board is not working properly. The authority has not paid attention to the solid waste pollution though the authorities of Pollution Control Board stay in Kolhapur.

Need for Environmental Education

In order to know awareness about the environmental problems the question was asked whether there was need for environmental education or not. The majority (92.8%) of heads of the households said that there is a need of creating environmental awareness and environmental education.

Housc Tax Paymcnt

The Kolhapur Municipal Corporation charges 'House Tax'. In order to know the payment of house tax paid by different households, the question was asked that how much amount was paid by them per annum. Out of 125 heads of the sample households, 68(54.4%)

households were paying less than Rs. 500.00 to KMC, 23(18.4%) sample households were paying in the range of Rs. 5001 to 1000, and 34(27.2%) sample households were paying more than Rs. 1000.

It was noticed in the field work that citizens were arguing that they were paying the House Taxes to KMC but they did not receive the solid waste management services properly. There is a special mention of 'sanitary charges' in the House Tax receipt but not used upto full extent. Out of total House Tax, about 33% of the amount was spent on solid waste management.

Willingness to Pay

In order to know the willingness to pay for private work of solid waste management, the question was asked to the heads of the sample households whether they would be ready to pay for private solid waste services. The response of the respondents is shown in Table 1.7. Out of 125 heads of the sample households, 118 (94.4%) heads of the households were ready to pay additional Rs. 50.00 per month for private agency. Hardly, 7 (5.6%) heads of the sample households said that they will pay more than Rs. 50.00 per month. It was also observed that people were arguing that as they were paying municipal tax to KMC, so they will not pay to private agency.

Suggestions for Better Solid Waste Management

For effective and scientific management, different heads of sample households had made different suggestions. These suggestions stands for :

1. Citizens should co-operate with the KMC by putting solid waste in bins.
2. People should involve in the management of solid waste services.
3. Privatisation of SWM services should be done.
4. Nuisances of rats, pigs and animals should be controlled.
5. Frequency of lifting of solid waste should be raised.

Table 1.7 : Willingness to Pay for Cleanliness

Sr. No.	*Name of the Locality*	*Rs. <50.00*	*Rs. >50.00*	*Total*
1.	The high standard of living locality like Nagala Park and Tarabal Park (having maximum number of bungalows and RCC type of housing units)	23 (92.0)	2 (8.0)	25 (100)
2.	The new extended area like Shahupuri and Rajarampuri (having maximum number of RCC and flat type of housing units)	20 (80.0)	5 (20.0)	25 (100)
3.	The Old Kolhapur like Mangalwar Peth and Shivaji Peth (having maximum number of old and Manglori roof type of housing units)	25 (100)	-	25 (100)
4.	The low standard of living area like Shahu Mill Workers Colony and Sidharthnagar (having maximum number of *kachha* and Manglori roof type of housing units)	25 (100)	-	25 (100)
5.	The slum area like Sadar Bazar and Rajendranagar (having maximum number of *kachha* and hut type of housing units)	25 (100)	-	25 (100)
	Total	**118 (94.0)**	**7 (5.6)**	**125 (100)**

Source : Field Survey.

6. KMC should make recycling of waste.
7. Workers strength should be increased.
8. NGOs should participate more effectively in the solid waste management services of KMC.

All the categories of people has suggested a need for change in the present disposal management of MSW; but they are not in favour to pay more municipal taxes or additional fees for this purpose. There are complaints against the municipal workers and officers engaged in this department, but the work sitution is so dirty that an ordinary citizen cannot stand even for a few minutes nearly such dust bins. While conducting the survey it was observed by the researcher that the fingers and feet of workers were having the mark of injuries while lifting the glass and other hazardous items in the bins.

REFERENCES

Aurther, H. Parcell, (1980), *The Waste Watches : A Citizen's Handbook for Conservation of Resources,* New York, Anchor Press, Doubleday, U.S.A.

Bhide, A.D, Sundaresan, B.B, (1974), 'Solid Waste Management in Developing Countries', Indian National Scientific Documentation Centre, New Delhi, India.

Jadhav, H.V., Bhosale, V.M. (1995), *Environmental Protection and Laws,* Himalaya Publishing House, Mumbai.

Jadhav, H.V., Bhosale, V.M. (1995), *Environmental Pollution*, Himalaya Publishing House, Mumbai.

Karanja, A.M., (1996), 'Alternative Approaches to Solid Waste Management in Nairobi: The Linkages between Formal and Informal Systems' (Dayster University, Nairobi, Kenya).

Karpagam, M. (1999), *Environmental Economics*, Sterling Publisher Pvt. Ltd., Delhi.

Mahamuni, V.V. (2002), Economic analysis of Solid Waste Management in Kolhapur Municipal Area: An environmental perspective Ph.D. Thesis submitted to Shivaji University, Kolhapur.

Mahbub, ul Haq, (1983), *The Poverty Curtain Choice for Third World*, Oxford University Press, Delhi.

Panjwani, P.U. (1998), Collection, Transportation and Disposal of Municipal Solid Waste in Bombay City, *Indian Environment Directory*, edited by Rashmi Mayur, Mumbai.

2

MUNICIPAL SOLID WASTE MANAGEMENT IN THANJAVUR

S. Padmavathi and *C. Sunitha*

Mahatma Gandhi, a man far ahead of his time, has rightly pointed out that nature has enough to provide for man's needs but can never provide for his greed. Reflecting Gandhiji's views. The Brundtland Report commissioned by U.N., called for the "sustainable development" that can meet the needs of the present without compromising the ability of future generations to meet their own needs.

Economic Development

The present "pseudo-development" of economy, depleting non-renewable resources and degrading renewable resources, may collapse the very foundation of development. Of course, when we clear forests, over harvest fisheries, outwork cropland until the soil erodes, then use our skies as free rubbish bins and our rivers as sewers, we can actually show an increase in our national income. But such strategies are shortsighted and such gains short-lived. (Agarwal & Dubey, 2002).

Almost all the developing nations were under the colonial rule till 1950s. After independence to achieve speedy growth, they borrowed the development models from the west, and yielded to industrialization and modernization, to increase their GNP, without ensuring equity, employment or minimum needs to their people. But, true development should be a process of utilization of environmental resources for the purpose of providing humanity with vital infrastructure facilities like education and health.

People as Cause and Victims

In Sir Edmund Hillary's words, environmental problems are really social problems; they begin with people as the cause and end with people as victims. So the growing environ-concerns are not of the governments and corporates alone. There is also a role for every individual to rethink his "right to pollute".

The preservation, protection and improvement of environment for present and future is the solemn duty of every citizen. The quality of his environment and quantity of waste he generates, are the best indicators to know whether a person performs this duty or not.

Agricultural wastes are mainly degradable. The public cannot control industrial wastes, which are mainly non-degradable. But the household solid wastes are controllable and widely avoidable by the people.

Today the disposal of garbage and other solid wastes appear to be much simpler. Urban dwellers merely place them in garbage cans and throw them out of their houses. But only when, one passes an open dump outside the city, the real magnitude of solid waste problem becomes apparent.

With the advent of "Use and Throw" concept, the solid waste has considerably increased which is a matter of growing concern particularly in urban India. So, to highlight the severity of these problems a micro level study is undertaken on the municipal solid waste management in Thanjavur Municipality.

Thanjavur Municipality

Thanjavur is a special grade municipality as well as District Head Quarters, located in the centre of Cauvery delta region. Being surrounded by core-wet lands with paddy crops Thanjavur is called the Rice Bowl of Tamil Nadu. It is very famous for the presence of Bragadeeswara Temple, which attracts a large number of foreign and domestic tourists. The city is situated at a distance of 351 kms away from Chennai, the capital of Tamil Nadu.

Profile of Tanjore Municipality

Area covered	36.31 sq.km.
Municipal population	1,97,635
Municipal population as in 2001	2,15,725
Mid-year population 2003	2,45,811
Number of families	49,026
Compost Area	20 acres

Solid Waste Management

The town generates 70 MT of solid waste at a rate of 310 gram per capita per day. But the present organized collection system can clear only 54 MT per day. The details are shown in the Table below.

Table 2.1 : Details of Solid Waste Management - Thanjavur Municipality

Solid Waste		*Quantity*
Total generation		70 MT
Collection/day		54 MT
% of coverage		77.14
Frequency of collection		Daily
No. of dust bins provided in the town		400
Per capita waste generation		310 gms
Per capita waste collection		240 gms
Disposal		
Disposal	Distance	Further usable period (years)
Site at Sekkadai Road	2 km from Town center	2 years

Source: Thanjavur Municipality

The Solid Waste collection coverage is around 77 per cent that leaves 23 per cent of the waste uncovered in the town. The wastes are disposed of at the Sekkadai road and no composting has been done for preparation of the manure.

Table 2.2 : Type of Solid Waste Collection—Thanjavur Municipality

Source	*Quantity (MT/per day)*	*Distribution (%)*
Household waste	18.4	34.10
Commercial and institutional waste	8.6	15.90
Market waste	15.9	29.50
Construction waste	11.1	20.50
Total	**54.0**	**100.00**

Source : Thanjavur Municipality.

The major proportion of solid waste consists of household, commercial and market waste, which amounts to 64 per cent of the total waste collected in a normal day. Normally market place generates sizable amount of waste per day in the town. The local body pays special attention to the market area at Karanthai, Gandhi market, Kamaraj market and Sarafoji market to clear the waste regularly. The construction waste constitutes 21 per cent of the total waste collected, which indicate increased construction activities in the local body.

The actual carrying capacity of the 33 vehicles in the Municipality is 4 MT. The vehicles can carry only 32 per cent of the total solid waste generated in a trip. Therefore, the vehicles are operating for 2 to 3 times in a day to clear the maximum solid waste accumulated in the town.

The proportion of road length to the existing conservancy staff works out to be 1289 metre per person, which is higher than the prescribed norms of 800-1000 metre person of the Central Government.

For the purpose of operational efficiency, the town is divided into 19 Zones and the Municipal Health Officer heads the Solid Waste Division. Sanitary Supervisors and Sanitary Inspectors assist him. There are nearly 208 sanitary workers for solid waste collection and handling in Thanjavur Town.

Short Falls Found in Solid Waste Management

The study revealed the following shortcomings in the present waste management system followed in Thanjavur Municipality.

1. With the available management mechanism only 77 per cent of the waste is collected and 23 per cent still remain unattended, which are likely to affect the health of the general public in this region. Lack of manpower, finance and coordination between different agencies is one of the reasons for the poor performance of the Thanjavur Municipality.
2. Plastic waste has been an important and increasing component of domestic refuse. This 138-year-old Municipality does not segregate it from the other degradable wastes so far. If these two are mixed, none of them can be re-used or recycled and the non-degradable waste will affect the decomposition process of degradable also.
3. All the wastes collected are just dumped in an uncontrolled and unsystematic manner. Open dumping at Sekkadai Road, without any safety precautions is a potential threat to the health of the people living in and around this area.
4. It is notable that the authorities do not implement even simple methods of composting. As the waste is not segregated from glass, plastics, metals and other hazardous substances, farmers and horticulturists are reluctant to use the compost as fertilizers. It leads to sizeable loss of income to the Municipality.
5. In Thanjavur Municipality alone there are 46 hospitals which are discharging huge amount of bandages, sanitary napkins and other infectious wastes, medicines, hair, glass bottles and pieces, plastic bottles, covers etc. Due to the absence of incineration facilities till December 2003, considerable amounts of these wastes were mixed and dumped together with other wastes that may create serious health hazards.
6. Besides bad odour and fly nuisance there is also a risk of toxic airborne gases from the dumping yard. Moreover, when rainwater percolates through this site, it decomposes the waste and liquid known as leachate is produced. If it is not properly treated, this toxic liquid may enter into the watercourses or ground water.

Eco-city Plan for Thanjavur

Under the 10th Plan, a budget provision of Rs.30 crores has been made for the period 2002-03 to 2006-07 for the Eco-city projects. Since Thanjavur is one of the important tourist spots, steps are taken to make it an Eco-city under this plan. "Eco-city" is a city that decreases environmental burden / stress, improves living conditions and helps in achieving sustainable development through a comprehensive urban improvement system involving planning and management of land and its resources and implementation of environmental measures.

The environmental improvement projects employed in the Eco-city project are innovative, to the non-conventional, practical and sustainable, cost effective, easily maintainable, simple in design, creating visible impact, replaceable, using locally available materials, adopting concepts and designs that suit the local people.

The important strategies to be followed under this project to make the city clean are:

- Achieving zero garbage mechanism by introducing private participation and door-to-door collection system.
- Strengthening the solid waste collection system through mechanized equipments and trained persons.
- Seeking co-operation of public towards controlling the non-degradable waste especially plastics of sizes ranging from 2 to 20 microns.
- Exploring the possibility of energy generation from the municipal solid waste.

Conclusion

Due to the changes in our economy, a change in the consumption pattern took place, in the last two decades. The increased quantity of the total waste is not only the outcome of population explosion but also because of the high per capita consumption, especially, of the products of marginal necessity and low product durability.

Solid waste is "a matter in the wrong place" implying that a material becomes waste only when a specific user ceases to have a use for it (Trivedi, 1997). In other words, it is a sign of NIMBY syndrome that is in-built in every individual. But the developing world with constraints of resources cannot afford to call any material as waste. Current waste disposal philosophy is to treat all wastes as resource materials. But the only requirement is that the segregation must precede the collection of wastes. Thanjavur Municipality finds difficult to manage the Municipal waste problem because of lack of segregation of wastes as compostable, recyclable and hazardous ones.

The present condition in both the global and local environment is critical to our health and well being. Indeed, environment related risk could be deadly; each year, infectious diseases linked to environmental conditions kill one out of every five children in the poorest region of the world. (Agro bios, Nov. 2003). So, keeping the city clean and green is the need of the hour.

Increased consumption and wastes generation can never be a sign of better standard of living. The development at the cost of environment is very myopic and we should not get indulged to go by this development, compromising on the environmental aspects, both at micro and macro level.

REFERENCE

Agro Bios vol.II No.6, Nov. 2003.

Annual Reports of Thanjavur Municipality.

Barry C. Field, *Environmental Economics*, McGraw Hill Companies, 1997.

Callan & Thomas, *Environmental Economics & Management,* Irwin Book Team, 1996.

Economic and Political Weekly, Nov.29, 2003.

G. Gaur, *Soil and Solid Waste Pollution and Its Management,* Sarup & Sons Publications, 1997.

M. Datta, B.P. Paride, B.K. Guha, T.R.Sreekrishnan, *Industrial solid waste management and land filling practice*, Narosa Publishing House, 1999.

Ministry of Environment & Forests, Municipal Solid Wastes (Management & Handling).

Pollution Research Enviro Media, vol 22(1), 2003

P.R.Trivedi & Gurdeep Raj, *Solid waste pollution*, Akashdeep Publishing House, 1997.

S.K. Agarwal & P.S. Dubey, *Environmental Controversies*, A.P.H. Publishing Corporation, 2002.

Sudhir Dawra, *Environmental Economics*, Mohit Publications, 2001.

Richard Wellford, *Corporate Environmental Management*, Universities Press, 1996.

Roma Mukherjee, *Environmental Economics in the Third Millenium*, Sterling Publishers, 2001.

TNPCB, Pollution Control Legislation.

V. S . Katiyar, *Environmental Concerns*, Pointer Publishers, 1997.

3

HEALTH IMPACT OF SOLID WASTE HANDLING

J. Rajesh Khanna

Introduction

Solid waste comprise of all the wastes arising from human and animal activities that are normally solid and they are discarded as useless or unwanted. The term solid waste includes the heterogeneous mass of throw away from the urban community as well as the more homogenous accumulation of the agricultural, industrial and mineral wastes. In early times, solid waste disposal was not a great problem as the population was small and mother nature was strong enough to assimilate the produced solid wastes. Problem started surfacing when humans first began to congregate in to tribes, villages and communities littering of food, and other solid wastes in medieval towns—led to the breeding of rats, with their attendant fleas carrying the bubonic plague. Lack of any plan for handling solid waste led to a number of epidermis and diseases resulting in loss of life of millions of people. In recent times, there has been a continuous exodus of people from rural and semi-urban areas to town and cities. In India population has expanded from 342 millions in 1947 to more than 1000 millions in 2000. Census 1991 shows that 220 millions out of 846 million Indians reside in about 4000 urban agglomerations. Though the entire country is facing severe problems due to increase in population and its ecological consequences, the impact under urban areas is manifold. The solid waste has become a life threatening issue for the society in urban local bodies, though committed to their services and finding is difficult to cope up due to the growing magnitude of this problem.

It is ultimate that lure of employment in urban areas and consequent urban migration is expected to continue and urban population would grow to 470 million by 2015AD; what is of great concern is that about 150 million of the urban population will be living in slums, thus adding more to *the owes of the urban local bodies* and to the society as a whole, the inability of the local bodies and lack of civic sense among the residents results improper handling and disposal of solid waste, which have an impact on the health of the public and environment. Public health authorities have shown that rats, flies, and other diseases vector breed in open dumps as well as in poorly maintained housing in food storage facilities and many other places where food and harborage are available for rats and insects associated with them. In recent study the (U.S.P.H.S.) US public health service has published the result of a study tracing the relationship of 22 human diseases to improper solid waste management.

The group at risk from the unscientific disposal of solid waste include the population in areas where there is no proper waste disposal method, especially the pre-school children; waste workers; and workers in facilities producing toxic and infectious material. Other high risk group includes population living close to a waster dump and those, whose water supply has become contaminated either due to waste dumping or leakage from landfill sites uncollected solid waste also increases risk of injury and infection. In particular organic domestic waste possesses a serious threat, since they ferment, creating conditions favourable to the survival and growth of microbial pathogens. Direct handling of solid waste can result in various types of infectious and chronic diseases with the waste workers and the rag pickers being the most vulnerable. Exposure to hazardous waste can affect human health, children being more vulnerable to these pollutants. In fact, direct exposure can lead to diseases through chemical exposure as the release of chemical waste into the environment leads to chemical poisoning. Many studies have been out in various parts of the world to establish a connection between health and hazardous waste. Waste from agriculture and industries can also cause serious health risks. Other than this, co-disposal of industrial hazardous waster with municipal waste can expose people to chemical and radioactive hazards. Uncollected solid waste can also obstruct storm water runoff, resulting in the forming of stagnant water bodies that become the breeding ground of disease. Waste dumped near a water source also causes contamination of the water body or the ground water source. Direct dumping of untreated waste in rivers, seas and lakes results in the plants and animals that feed on it. Disposal of hospital and other medical waste requires special attention since this can create major health hazards. This waste generated from the hospitals, health care centres, medical laboratories and research centres such as discarded syringe needles, bandages, swabs, plasters and other types of infectious waste are often disposed with the regular non-infectious waste. Waste treatment and disposal sites can also create health hazards for the neighbourhood. Improperly operated incineration plants cause air pollution and improperly managed and designed landfills attract all types of insects and rodents that spread disease. Ideally these sites should be located at a safe distance from all human settlement. Landfill sites should be well lined and walled to ensure that there is no leakage into the nearby ground water sources. Recycling too carries health risks if proper precautions are not taken. Workers working with waste containing chemical and metals may experience toxic exposure. Disposal of health care wastes require special attention since it can create major health hazards, such as hepatitis B and C, through wounds caused by discarded syringes. Rag pickers and other, who are involved in scavenging in the waste dumps for items that can be recycled, may sustain injuries and come into direct contact with these infectious items.

Classification of Solid Waste

Solid waste includes both solid and semi-solid waste that the possessor no longer considers of sufficient value to retain. A number of classification of solid wastes have been developed, and the following categories are useful :

- Resident
- Commercial
- Institutional
- Construction and demolition
- Municipal services
- Treatment plant sites waste
- Industrial
- Agricultural

Hazardous Solid Waste

They may be classified as wastes of industrial, institutional consumer origin, which, because of their physical, chemical or biological characteristics are potentially dangerous to human and the environment. Examples are solvent, paints, pesticides, and pathological wastes from hospitals and from research laboratories, which often find their way in municipal waste and become part of Urban waste stream. Though every kind of waste to some extent has deleterious effect on public and environment, but hazardous waste is the one, which has immediate effect on the people who come in physical contact and on the environment. In other words, hazardous waste can be defined us waste or combination of wastes that pose substantial present or potential hazard to human or other living organisms because:

- Such wastes are non-degradable or persistent in nature,
- They can be biologically magnified,
- They can be lethal in nature, or
- They may otherwise cause or tend to cause detrimental cumulative effects.

Properties that are used as a measure to classify waste as hazardous are : Corrosivity, Exokisivity, Flammability, Ignitability and Reactivity. While health related properties are Infectivity, Carcinogrinity, Irritant (allergic response), Mutagenity, Toxicity Chronic, acute) Radioactivity, Teratogenicity. When dealing with

MSW the materials found are classified as hazardous according to the following: Ignitability, Corrosivity, Reactivity, Toxicity, Carcinogenity. Hazards due to various types of solid waste can be summarized as :

(a) Hazards Due to Residential Waste

The residential waste generally consists of discarded products like paints, insecticides pesticides, automotive products, medicines, alkaline batteries, electrical equipments, etc. These products are highly hazardous owing to their corrosive flammable, irritant and poisonous properties. Paints, insecticides and pesticides though found in small quantities are persistent in nature, which is a critical issue for their long-term management. They enter the food chain resulting in bio magnifications hence being dangerous to mankind. They cause toxicity problems primarily to environment and biota at the source or point of release. Toxic effect occurs rapidly after exposure, (acute and sub-acute). The transport of these organic wastes (like chlorinated insecticides, pesticides, etc.) from the source results in widespread contamination and bio-concentration in the food chain, thus endangering the lives of the people at large. Environmental transport may expose biota to lower level of the pollutant, resulting in chronic toxicity. Automotive products, electrical equipments, household, batteries contain lead, mercury, arsenic, copper, chromium, cadmium, barium, etc. These elements make the waste carcinogenic and mutagenic. It may cause red staining of fingers, teeth, hair, general weakness, irritation of mouth and nose, dermatitis and fatigue among those who come in direct contact with the above mentioned waste. When such waste is disposed of in the landfills it may contaminate the soil, surface and ground water through the lechate, which is rich in these elements. The ingestion may cause damage to central nervous system, and may concentrate in pancreas, thyroid. They biomagnifies leading to impair of reproduction and in even death. Household batteries are the main culprit behind the contamination of underground water by Arsenic and Mercury. These make the waste highly toxic to handle as these elements can be absorbed by skin and inhaled. They lead to birth defects loss of memory dermatitis and even in severe cases death. They are health deleterious to natural flora/ fauna.

(b) Hazards Due to Industrial Waste

Very few studies have been conducted on health hazards due to industrial waste. There have been reports that sacks, cardboards and paper envelopes contaminated with chemicals when burned produced irritating fumes that caused respiratory problems and exposure to such chemicals led to skin disorders, etc.

(c) Hazards Due to Commercial Waste

The hazardous waste produced by commercial waste depends upon the services provided. Examples are inks from print shops, solvents from dry cleaning, petroleum based products from painting contractors, etc. The dry cleaning petroleum based products from painting contractors, etc. the waste can be highly explosive, corrosive and ignitable thus, endangering the lives of the personnel handling these.

(d) Hazards Due to Institutional Waste

This waste varies widely according to the services being offered. The major concern is the bio-medical waste which is resulting from hospitals and research laboratories. The waste contain needles, bandages, discarded samples (blood, urine, etc.), and materials that are highly infected of pathogens. Such kind of waste may result in an increase in disease like cholera, plague, AIDS, hepatitis, diphtheria, etc., in either endemic or epidemic form, posing great risk to public at large. This may generally consist of insecticides, fumigants, pesticides, herbicides, etc. these products contain organic compounds like silvex, lindane, toxaphene, etc., these materials are highly toxic when inhaled, ingested or when someone comes in direct contact as the skin absorbs them.

Health Hazards Due to Handling of Solid Waste

Physical contact with solid waste has long been regarded as a potential public health risk but nothing much has been done for those who come in direct contact of the solid waste, while assumptions and study concerning the solid waste disposal has served as basis for promulgation for quality of environment. The most obvious detrimental effect due to lack of proper handling of solid waste is aesthetic, ugliness of street litter city. More serious is the transfer of pollution to water, both surface and ground. It also leads to air pollution due to inefficient burning of waste. Uncontrolled hazardous wastes from industries and institutions, which often find their way in municipal waste, may create havoc to the environment and health of the people. These are specific changes of concentration of heavy metals and other toxic compounds in food chain due to seepage in ground or illegal disposal of industrial effluent. Though there is no well documented and direct evidence to indicate or suggest that specific risks are encountered through exposure to solid waste, but epidemiological studies have shown that high percentage of workers who handle refuse and of individuals who live near or on disposal sites, are organisms. Contamination of this kind is likely at all points where waste is handled. Although it is known that vector insects and rodents can transmit various pathogenic agents (amoebic and bacillary dysentery, typhoid, salmonellas, cholera, yellow fever, leptospirosis, skin diseases, AIDS, etc.), it is often difficult to trace the effects of such transmission to a specific population.

Even then, the study on the health hazards encountered by workers handling refuse in none the less important as it can be used as and indicator of the public health risk due to the physical contact with the solid waste and well as due to indirect consequences of improper solid waste management such studies are yet to undertaken and given due recognition if environmental clean up efforts are to be adequate to safeguard public health. Unless solid waste is appropriately managed, its adverse impact will continue until it has fully decomposed or otherwise stabilized. Most development activities are expected to have beneficial effect on human health by increasing resources available for food, education, and employment, water supply, sanitation and health services. Proper management of MSW should have minimum effects on environment and on public health. Clark published the most extensive investigations of disease risks due to occupational exposure in year 1976. The study was conducted through medical history questionnaire surveys and clinical evaluations of workers health. The refuse workers were found to be suffering from disease like leptospirosis also known as weil's disease. The workers and 1 per cent of the street sweepers suffered from this disease. The workers were also prone to infectious hepatitis due to unavailability of clean drinking water and other unsanitary conditions prevailing at the dump sites. Also the solid waste disposal sites lacked first aid medical help thus increasing the risk many times. It was found through the study that the transmission of by usual means of person-to-person contact is considerably more likely as a case of infection. Therefore, no incidence rate could be determined. A rise in skin disease was found in people working at the processing and disposal of solid waste suffered from eczema and intermycoses, often after the commencement of the employment. In other study done by Englehardt for the Florida center for solid and hazardous waste management (using survey of 251 garbage collectors and 4 supervisors in Florida) indicated high rates of mortality and injury to drivers and helpers involved in collection and transportation of MSW. An average of 9.8 workers compensation arise involving greater than seven calendar days lost sork per 100 workers per year, was found for garbage collectors. This rate is 7.4 times the rate for the general Florida workforce. Mortality for MSW handling drivers and helpers were estimated at round 90 deaths per 100000 workers per year ranking in between rates reported for the second and the third deadlist occupations (timber cutters and air plane pilots) nationally. Actual injuries were found to be an order of magnitude higher that the number of workers compensation claims.

CPHERI in a study estimates that 64 per cent of refuse workers suffered from hookworm and 68 per cent from round worm. A summary of health status is given below :

In metropolitan and urban areas scavenging is an important activity and an important unorganized private industry employing thousands of scavengers who live on refuse dumps. The estimated numbers of rag pickers in few cities are :

City	*Number*
Mumbai	1,00,000
Delhi	75,000
Kolkata	40,000
Ahmedabad	30,000
Chennai	30,000
Bangalore	25,000
Jaipur	10,000

Many of them from nearby villages and are lured into this profession by those involved in recycling of waste. They may also be the abandoned children and those from destitute families residing in the slums of the cities. They live under health risks and suffer severe exploitation. The ailments suffered by rag pickers are quite similar to the *safai karamcharis* who are involved in primary collection of waste. The possible health hazards are contusion, laceration, gastrointestinal problems, eye infections, lower back pain, skin disorders, malnutrition etc.

Conclusion

In recent times because of exponential rise in the amount of solid waste generation it has become a great concern for the society as a whole. Although there is a dire need to expand the horizon of muicipal services but more importantly there is a need to educate society at large to develop right practices of solid waste disposal, particularly relating to hazardous industrial and biomedical waste.

society at large to develop right practices of solid waste disposal, particularly relating to hazardous industrial and biomedical waste. Solid waste workers are exposed to significant levels of physical, chemical and biological toxins. Injuries as well as acute and chronic muscular skeletal, demal and respiratory health effects are well documented among these workers. Engineering controls, exposure monitoring, education, personal protection, and other interventions are under-utilized in protecting solid waste from exposure and health effects. The section which is bearing the burnt of this rise in quantum of solid waste are those that are engaged in its disposal and thus Government should take the onus along with NGOs to help this strata of the society which due to ignorance and lack of education are not aware of the potential risks associated with handling of such waste. Workshops, road shows, campaigns, should be organized to create awareness about solid waste handling. Along with other cases of occupational health, hazards of solid waste should be well documented for developing countries like India and government should work upon bringing a legislation to regulate the health and safety of workers involved in solid waste handling.

REFERENCES

Howard S.Peavey, Donald R.Rowe, Georage *Technobanglous Environmental Engineering*, International edition, MC-graw Hill, 1985.

James Englehardt, *Florida Municipal Solid Waste Collection : Occupation Risks and Prevention*, www.Florida.org.

Manual of municipal solid waste management first, edition May, 2000, pp. 454-458.

Technobanoglous, Hilary Theisen, Samuel & Vigil, *Integrated solid waste management engineering principle and management issues*, International Edition, 1993,p.5.

4

SOLID WASTE MANAGEMENT IN SMALL TOWN

A Case of Chinnalapatty

P. Ayyammal, S. Ramaswamy and *G. Sathis Kumar*

Introduction

Solid waste management accounts for the highest proportion of all types of wastes. Management of burgeoning solid wastes becomes a critical for almost all the major cities in India. Rapid population growth, industrial development and solid waste disposal have led to severe problem of waste management before civic authorities of the cities. Solid waste management is a integral part of urban and environmental management of each city with more than 65 per cent of India's 250 million population living in class one towns [population over one lakh] and 23 cities getting distinction of being metropolises [population over one million]. Solid waste management was never taken up seriously, either by the public or by concerned agencies or authorities and now the piled up waste is threatening our health, environment and well-being. The present paper focuses on solid waste generation by different economic sectors of Chinnalapatty economy.

Conceptualization

Solid : A state of matter in which the relative motion of the molecules is restricted and they lend to retain a definite fixed position relative to each other, giving rise to crystal structure. A solid may be said to have definite shape and volume.

Solid waste: Anything solid discarded for any reason, whether part or all of the substance will be recycled—rubbish, trash, scarps, fecal matter, harvest leavings, sewage sludge, bottles, cans, tires, paper, chemicals, slag, used cars and machinery, mining tailings scarp metal.

Solid waste pollution: An apt definition of solid waste is "matter in the wrong place"; current waste disposal philosophy is to a endeavour to treat all wastes as resource material, some for recycling, some for conversion to fertilizer or as a source of energy, and the balance of reclamation.

Solid waste management: A planned system of effectively controlling the production, storage, collection, transportation, processing and disposal or utilization of solid waste in a sanitary, aesthetically acceptable, and economic manner. It includes all administrative, financial, legal and planning function, as well as the physical aspects of solid waste handling.

Effective Group of Wastes

The responsibility of solid waste management remains primarily with municipal bodies; several other stakeholder groups also play significant roles in the solid waste management. It is widely recognized that wastes can be effectively grouped according to Origin, Properties, Legal Definition, Health risk, and its frontiers. There is no need to describe the groups, because they are self-explanatory.

Origin of Solid Wastes	*Forms of Solid Wastes*
Agriculture wastes	Liquid/ Gels
Household or Municipal Wastes	Solid
Clinical or Bio-medical Wastes	Gaseous
Industrial Wastes	Slurries
Radioactive Wastes	Powder

Properties of Solid Wastes		*Legal Definitions of Solid Wastes*
Toxic	Inflammable	Household
Hazardous	Carcinogenic	Industrial
Tactile	Volatile	Controlled
Acidic		Directive
Alkaline		Hazardous
Inert		Special

Health Risks of Solid Wastes

Infections
Inflammable or Explosive
Hazardous Chemicals
Radioactive Chemicals

Objectives of the Present Study

Major objectives of this study are:

- Tofind out the sources of waste generated in different economic sectors of Chinnalapatty.
- To estimate the volume of waste generated in the study area.
- To find out the problems, if any, in solid waste management process and suggest suitable, and scientific mechanisms in managing the solid wastes.

Solid Waste: Indian Scenario

It is well known fact that, there is a close relationship between GNP per capita and per capita waste generation. Since, the solid waste quantities are directly proportionate to the quantity of material consumed, the increase in per capita solid waste quantities would be directly proportional to the per capita increase in GNP. This is statistically shown in the Table 4.1.

Table 4.1: Relations between GNP and Expected Generation of Municipal Solid Waste

Economies	*1995*		*2025*	
	GNP per capita [US $]	*Urban Waste Generation [kg/capita/day]*	*GNP per capita [US $]*	*Urban Waste Generation [kg/capita/day]*
Low Income	490	0.64	1,050	0.6-1.0
Middle Income	1,410	0.73	3,390	0.8-1.5
High Income	30,990	1.64	41,140	1.1-4.5

Source : Government of India: *Manual on Municipal Solid Waste Management* (Ministry of Urban Development, New Delhi, May 2000).

Due importance has not been given to the subject of solid waste management in India. On account of low priority given, the solid waste management practices have continued to remain inefficient and outdated. Lack of national policy, lack of interest by the State Government, inability of the local bodies to provide efficient services, lack of awareness and apathy of the community towards solid waste management, lack of involvement of NGOs and use of inappropriate technology for collection of transportation and disposal of waste, institutional weakness, lack of trained manpower, lack of committed financial resources, lack of legislative support and ineffective implementation of laws and regulations are responsible for ineffective SWM services in the country [P.U.Asnani,1996].

The volume of garbage in Indian cities is increasing. Indian cities and towns are estimated to generate about 80,000 metric tonnes of solid waste everyday. Per capita solid waste generated is about 350-400 gms and in large cities it exceeds 500 gms. Only 60 per cent of this volume is collected, even less is transported and disposed of. Sanitary landfill or composting as methods of garbage disposal is limited to very few cities. Mumbai generates about 3200 tonnes of garbage of which about 97 per cent is collected. In other cities, collection varies from about 90 per cent in Delhi and Chennai, about 75 per cent in Calcutta and Hyderabad, 68 per cent in Bangalore and 70 per cent in Kanpur. The collection is around 50 per cent in smaller towns. In industrial areas of many cities, the municipal solid waste is getting mixed up with hazardous waste creating serious environmental problems, while the accumulation of garbage has become a common site in most of the cities. Most solid wastes that are collected end up in open dumps, sanitary landfill or drainage system, threatening both surface water and ground water quality. Solid wastes create one of the most visible environmental problems in low income areas. These problems are directly linked to inadequate planning, finances and management capacity at the local level. [K.C. Sivaramakrishnan and B. N. Singh]

Methodology

The study is an empirical and analytical in nature, based on the primary and secondary data. The primary data have been collected from different economic sectors, which creates solid wastes for which separate interview schedules were prepared and pre-tested. The pre-tested interview schedules were used for collecting data and information from different economic sectors, viz. Households, Hospitals, Markets, Marriage Halls, Schools, Hostels, Tailoring Shops, Printing Press and Hotels.

Profile of the Study Area

The area selected for this study is Chinnalapatty—a small town, situated very nearer to Gandhigram. It is a town panchayat located in Dindigul District, Tamil Nadu, India. It is 13 kms away from Dindigul and 47 kms away from Madurai on NH 7. It is well connected by road and railways. Most of the people are involved in traditional weaving and its connected activities. It consists of 16 wards with 6500 households. It possesses sufficient socio-economic infrastructural facilities like educational institutions, entertainment and commercial centres, communication, hospitals, government offices etc.

Solid Waste—Different Views

The following are the different views expressed by the various experts who have done different research studies on management of wastes in different time periods.

- Rapid growth of population, industrialization, urbanization, commercialization, and consumerism have led to the migration of people from rural to the industrial cities, causing accumulation of solid wastes. [Singh and Singh, 1998]
- The accumulation and improper disposal of wastes leads to environmental pollution and accelerates the spread of communicable diseases. [Bhat Sairam, 2002]
- Organic wastes can either be recycled as animal feed or be easily disposal off in village dumps, which do not require a significant amount of public intervention. Non-organic wastes can be either burnt or dumped and buried again without need for public intervention or environmental damage. [Johannes, 1981]
- Composting municipal solid waste results in recycling of the waste and also generated compost which can be used as manure. [Alone and Bhide, 2002]
- Vermiculture technology can be used effectively for composting of urban wastes and thus the disposal problems of urban wastes can be handled effectively. [Jeevan Rao and Rama Laksmi, 2002]
- Need for organizing workshops, road shows, campaigns in order to create awareness about solid waste handling. [Sheeba Nishat, 2002]

Micro-level Evidence

The researcher have identified different wastes generating sectors in Chinnalapatty town. Major wastes generating sectors are carefully analyzed and estimated the quantity of wastes generated by each sector in the study area, presented in the table and diagram, which are self-explanatory.

Table 4.2 : Annual Generation of Wastes from Different Economic Sectors in Chinnalapatty

Sl. No.	*Waste Generating Economic Sectors*	*No. of Respondents*	*Total waste [in tones]*	*Percentage*
1.	Households*	6500	2804.6	76.0
2.	Hospitals	4	680.9	18.4
3.	Markets	25	109.8	3.0
4.	Marriage Halls	10	54.5	1.4
5.	Hotels	20	18.4	0.5
6.	Industries	4	8.3	0.2
7.	Schools	4	6.0	0.2
8.	Hostels	2	4.6	0.1
9.	Tailoring Shops	27	3.8	0.1
10.	Printing Press	3	2.0	0.1
	Total	174	3692.9	100.0

Note: *Wastes generated from the households have been estimated, based on a sample of 75 households were contacted by the researcher through a pre-tested interview schedule.

Source: Computed from the primary data.

Fig. 4.1 Annual Generation of Wastes from different Economic Sectors in Chinnalapatty

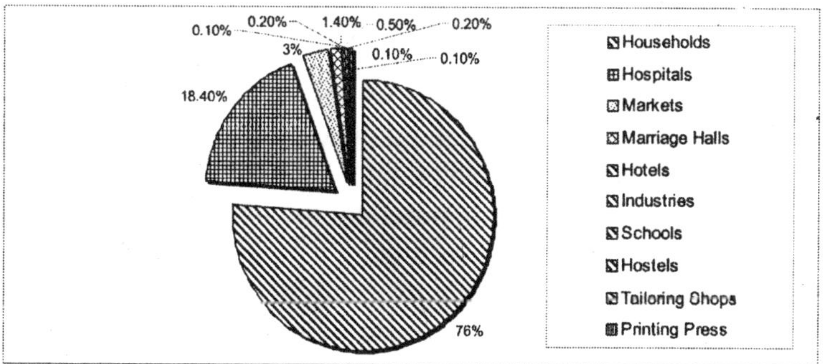

Major Findings

- The quantity of wastes generated in the study area is 3692.9 tonnes per annum. Of which, household sector generates

76 per cent of wastes following by hospitals [18.4 per cent], markets [3.0 per cent], marriage halls [1.4 per cent], hotels [0.5 per cent], industries [0.2 per cent], schools [0.2 per cent], hostels [0.1 per cent], tailoring shops [0.1 per cent] and printing press [0.1 per cent];

- Proper scientific mechanisms are not available in disposing the wastes generated by the different economic sectors in the study area; and
- People in general and economic sectors in particular never bother about the neighbourhood environment.

Suggestions

The major suggestions emerged out of this study are:

- Creating awareness among the public about effective utilization of wastes;
- Informing the people about the conversion of wastes into wealth;
- Providing minimal infrastructure facilities for collecting the wastes from different economic sectors;
- Imposing tax on public, especially waste generating sectors, for ineffective management of wastes;
- Providing incentives for those sectors which adopt better and effective handling of wastes;
- Honouring the, households who keep the surroundings of the house in neat and clean;
- Waste tax can be imposed by the local government on different economic sectors.
- Hospitals should have two separate sets of arrangements for storage of hospital waste—One for Bio-medical waste and the other for the ordinary municipal waste.
- Establish an institution of waste management at the national level to guide the urban area.
- Markets and hotels produce huge quantity of organic waste This waste is ideal for composting. Such waste should be

deposited directly in large containers which should be taken to the compost plant or disposal site.

- Large communal waste storage bins for commercial areas as well as for residential areas as on alternative to community bins.

Conclusion

Management of wastes is an important problem faced by the local governments, like Panchayats, Municipalities and Corporations. Each and everyday, large quantities of wastes are being generated by the different economic sectors. These wastes could be effectively managed and in some cases effectively conserved. Effective management and conservation of wastes generally conserve the natural as wall as environmental resources. Further, conservation of natural resources and effective management of wastes can reduce the problems of pollution and thus save man and nature. Natural Environment can be protected by adopting appropriate wastes management policies, which should be effectively implemented by the authorities. The success of waste management schemes and programmes depends upon the coordination of the different government machineries with support of effective people's participation.

REFERENCES

Alone, B.Z. and Bhide, A.D.: "Standards for compost from Municipal Solid Waste", Solid waste management—Current Status and Strategies for Future, 12-14, December 2002

Barthhold A. Thomas: "Issues in Design of Environmental Excise Taxes" *Journal of Economic Perspective* (8), 1994, pp. 133-145.

Basu, S.: Medical Waste Disposal Burning Problem, *The Hindu Survey of Environment*, Madras 1998, pp. 25-30.

Bhat Shairam : "Legal Response to Manage the Challenges of 'Wastes' in India", Solid Waste Management—Current Status and Strategies for Future, 12-14, December 2002.

Choudhari A. Singh, P. Kshirsager, D.G., Wate, S. R and Sarin, R.: "Hazardous Waste Management at Petrochemical Industry", *Indian Journal of Environmental Protection.*Vol. 22, No. 8, August 2002, pp. 858-862.

Government of India : Manual on Municipal Solid Waste Management, The Expert Committee: Central Publication Health and Environmental Engineering Organization (CPHEEO), Ministry of Urban Development, New Delhi, 2000.

Jeevan Rao and Rama Lakshmi, Ch. S : "Vermiculture Technology for Effective UrbanWaste Management" Solid Waste Management—Current Status and Strategies for Future, 12-14, December 2002.

Municipal Solid Waste: www.epa.gov/cpaoswer/non-hm/muncpl/index.htm

Plourde and Smith,: "Environmental Economics", New Delhi, 1972, pp. 101-210.

Ramachandra, T.V. and Shruthi, B.V.: "Role of GIS in Solid Waste Planning and Management". Paper Presented in the 3rd International Conference on Environment and Health, 15-17 December 2003. Chennai, pp. 66-67.

Ruth A. Eblen and William.: "Municipal Solid Waste Management", *The Encyclopaedia of the Environment.* New York, 1994, pp. 793-795.

Sheepa Nishat : "A Perspective on Health Hazards of Solid Waste Handling", Solid Waste Management—Current Status and Strategies for Future, 12 -14 December 2002.

Singh, S.K. and Singh, R.S. : "A Study on Municipal Solid Waste and its Management Practices in Dhandadjharia Coalfield". *Indian Journal of Environmental Protection*, Vol. II, No. 11, November 1998 pp.850-852.

Trivedi, P.R. : "International Encyclopedia of Ecology and Environment"— Solid Waste Management, Vol. 17, Indian Institute of Ecology and Environment, New Delhi,1995.

WHO: Waste from Health—are activities WHO Information, Fact sheet N-253 October 2003. http.www.int/inf.fs/en/fact 253. html.

5

ENVIRONMENTAL IMPACT OF SOLID WASTE INCINERATION

V. Dhulasi Birundha and *R. Meena Devi*

Introduction

The process of burning of solid waste at high temperature in large furnaces is known as incineration. In these plants the recyclable material is segregated and the rest of the material is burnt. At the end of the process all that is left behind is ash. During the process some of the ash floats out with the hot air. This is called flyash. Both the flyash and the ash that is left in the furnace after burning have high concentrations of dangerous toxins such as dioxins and heavy metals. Disposing of this ash is a problem. The ash that is buried at the landfills leaches the area and cause severe contamination (http: //edugreen.teri.risin/explore/solwasts/ disposal.htm)

Existence of solid waste is harmful to the environment, and may cause effects on public health and aesthetics to affect both natural and living environment. These effects are the primary environmental impact. Their storage, collection, transport, treatment, disposal and final return to the natural environment produce secondary environmental impact. For example: air pollution may result from incineration and offensive odours, and water contamination may be caused by the landfill disposal of solid wastes. (Hee-jung Park, 1992)

Incineration of Solid Waste

Industrial solid wastes that are incinerated are mostly cellulose type and more often industries have to handle certain kinds of chemical wastes in the form of either solids or sludges. The advantages of incineration include wide variety ability for handling varying loads and small space requirement for ultimate disposal. However, the method requires fairly high level of maintenance and the operating costs are higher than those for the operation of a sanitary landfill. (C.S. Rao)

Incineration systems use high-temperature combustion under controlled conditions to convert wastes containing infectious and pathological material to inert mineral residues and gases. The three types of incinerators used for hospital waste treatment are multiple-hearth, rotary kiln and controlled air. All these types can have primary and secondary combustion chambers to ensure maximum combustion of the waste. (Parivesh, 1998)

Comparison of Treatment Technologies

All the treatment technologies explained above have their own advantages and disadvantages. Table 5.1 overleaf, gives the relative advantages of the treatment technologies :

EMISSIONS FROM MEDICAL WASTE INCINERATION

Particulate Matter

Particulate matter is generated when non-combustible material is suspended, when incomplete combustible of combustible materials occurs and the vaporous materials condense.

Toxic Metals

Toxic metals appear in emissions as particulates. The concentration of metal emissions depends on the quantity of metals in the waste material. The metals of concern which are expected to be emitted are arsenic, cadmium, led, mercury, etc.

Talble 5.1 : Overleaf, gives the relative advantages of the treatment technologies

Treatment system	*Autoclave*	*Hydroclave*	*Microclave*	*Incinerator*	*Chemical*
Capital cost	Low	Low	High	High	Moderate
Operating cost	Low	High	High	High	Low
Operator maintenance skills	Low skill level required	Low skill level required	Automated but highly complex and high level maintenance skill required.	High level operator and maintenance skilled require	High level required for chemical control and grinder
Air emissions	Odorous but non-toxic	Some what odorous but non-toxic	Some what odorous but non-toxic	Can be highly toxic	Some chlorine emissions
Water emissions	Odorous may contain live micro organisms	Odorous but sterile	Negligible	None	None
Treated waste characteristics	Wet waste all material recognizable	Dehydrated, shredded waste, unrecognizable material	Shredded but wet waste	Mostly ash, may contain toxic substances	Shredded wet waste, containing chemicals used as disinfectants

Source: Parvish (1993).

Toxic Organics

Toxic organics can be formed of incomplete combustion. For instance, chlorine, which is derived from the incineration of PVC plastics, can combine with organics to form toxic chlorinated organics, such as dioxins and furans, which are best, avoided.

Carbon Monoxide

Carbon monoxide is a product of incomplete combustion and is a good measure of incinerator efficiency. Carbon monoxide production is limited when oxygen concentrations, mixing, and temperature are adequate.

Acid Gases

Acid gases (hydrogen chloride, sulphur dioxide, nitrous oxide) are formed when nitrogen, sulphur, and chlorine are released during combustion. Atmospheric nitrogen also combines with oxygen to produce nitrogen oxides.

ENVIRONMENTAL IMPACTS OF SOLID WASTE INCINERATION

Carbon Monoxide

Carbon monoxide (CO) is produced when oxygen provided is insufficient for a complete combustion of the waste. It is, therefore, an indicative of burning, or combustion efficacy. (Baranek,1971). The generation of CO decreases with an increase in the air supply, but increases with an increase in temperature. Carbon monoxide, when inhaled, passes through the lungs and diffuses directly into the blood stream where it combines with the red blood pigment called haemoglobin forming carboxyhaemoglobin, COHb. The affinity of carbon monoxide for haemoglobin is 210 times greater than that of oxygen for body tissue is considerably reduced. They induce heart and respiratory disorders. (Stewart, 1975)

Nitrogen Oxide

The amount of nitrogen oxides generated is a function of temperature, excess air, refuse composition, etc. The greater the amount of excess air, the greater the amount of nitrogen, and higher quantities of nitrous oxide would be expected. (Burnner, 1983) No is not an irritant and at concentrations that occur in the atmosphere it does not show any adverse health effects. However, its main toxic potential results from its oxidation to nitrogen dioxide. Nitrogen dioxide is relatively insoluble and upon inhalation can reach the moisture-filled alveoli of the lungs.

Pollutant	*Level (ppm) and Exposure*	*Effects*
Nitrogen dioxide	0.25for 8 months	Increased Obscession and reduced yield in citrus plants
	0.5 for 10-12 days	Suppressed growth of tomatoes
	3.5 for 21 hours	Spots of mild necrosis on cotton and bean plants
	25 for 1 hour	Acute leaf injury

Source : USEPA-1971.

Sulphur Oxide

Sulphur contained in solid waste is released into the atmosphere during the burning process, in the form of sulphur dioxide and sulphur trioxide. More than 95 per cent of sulphur oxides generated are sulphur dioxide, which will be slowly oxidized to sulphur trioxide. It is difficult to predict the quantities of sulphur dioxide and sulphur trioxide generated during incineration. However, many wastes have low sulphur content and thus the release of sulphur oxides is not significant. There is considerable evidence that sulphur dioxide aggravates existing respiratory diseases in humans and contributes to their development. Even healthy individuals experience broncho-constriction when exposed for a few minutes to levels of 1.6 ppm. (Stoker & Seager 1976) On vegetation it causes injury to spinach, and other leafy vegetables. Chlorosis and bleached spots are some of the symptoms. The most notorious pollutant responsible for metallic corrosion is sulphur dioxide. Sulphuric acid mixed in the atmosphere causes deterioration of structural materials such as marble and limestone.

Suspended Particulates

The most common particulate discharges come from the burning process as smoke. Smoke is a suspension of solid or liquid particulate matter with in a gaseous discharge. The particles sizes range from a fractions of a micrometer to over 50 micros. The visibility of smoke is related to the number of particles present, rather than the weight of the particulate matter (Burnner, 1983).

Hydrogen Chloride

Hydrogen chloride (HCL) is produced from the chlorine contained in refuse, which is exhausted with the combustion gas. It is a hazardous and corrosive. In refuse, chloride is contained at high ratios in plastics, such as PVC (Hyrayama, 1977)

Heavy Metals

Municipal solid waste incinerators produce more heavy metals than from other solid fuel fired plants, for example, coal, peat and wood waste (Carisson, 1986). The waste components contain varying quantities of metals and acid forming elements in municipal solid waste incinerators, most of the heavy metals (ZN, PB,CU,NI etc.,), remain in the bottom ash after burning of refuse. But, mercury is almost completely evaporated at the common combustion temperature of 800-1000° C.

Dioxin Emission Inventory

Dioxins (poly-chlorinated di-benzo-p-dioxins, or PCDD) and furans (poly-chlorinated-di-benzo furans, or PCDF) have been found in the environment. Combustion, especially of organics in the presence of chlorine in some form, has been found to be a significant source of these very toxic compounds (ASME, 1984). Dioxins interferes the normal expression genes, stimulates enzymes, and alters the metabolism of hormones, growth factors. At low exposure level it affects the thyroid gland. It causes skin trash, weight loss, fatigue altered glucose metabolism, neurological damage.

Dioxin Emission Inventory

MSW incineration	-	0.2 mg/kg
Hospital incineration	-	0.35 mg/kg
Apartment incineration	-	1.0 mg/kg

Waste Water

The waste water generated from refuse incineration plants is often regarded as problematic waste-water, because in many cases it contains higher pollutant and heavy metal concentrations than are permissible by the regulating authorities. Such contamination can be attributed to incineration of the heterogeneous products of refuse, with its multitude of pollutants entering the waste-water via various cleaning and cooling processes. The main pollutants result from removable inorganic substances, heavy metal ions, salts, temperature, and *p*H. One tonne of refuse incineration will produce 0.5 to $3m^3$ of waste water or about $3m^3$ when wet type dust collectors are used. (JICA,1983)

Noise

Among the various noise sources in an incineration plant, fans for blowing air are typical source. The noise level at plant boundary is important due to its impact on its adjacent areas. (Beranek, 1971)

Transportation

Construction of incineration plants results in a concentration of colletction vehicles to the vicinity of the plant site. The transporatation of collection vehicles also causes aditional traffic noise and air pollution on roadside environments. Sometimes too many vehicles are concentrated around a plant site, resulting in serious traffic problems.

Conclusion

From the environmental impact point of view, incineration causes drastic reductions in the volume of the refuse and the resulting residue is rendered largely inert and is, therefore, ideal for land filling. However, incineration of waste gives rise to a number of atmospheric emissions, which would have to be removed to prevent adverse environmental impacts. Suitable air pollution control devices shall be installed to meet the above mentioned emission standards. For setting up of incinerator, necessary clearance/consent shall be obtained from the concerned state pollution control board/ committee. Special care must be taken while transporting the wastes from the hospital to the incinerator in closed containers. The protocol for transportation shall be finalized by the local authority in consultation with the state pollution control board/committee.

REFERENCES

ASME, 1984, National waste processing conference. The American Society of Mechanical Engineers, New York.

Beranek, L.L.,(1971) *Noise and Vibration Control*, Mc.Graw Hill Book Co., New York.

Brunner, C.R., (1983) *Incineration Systems,* Government institute, Inc., Rockvile, MD.

Carisson, K. 1986., Heavy metals form energy from waste plants comparison of gas cleaning system. Waste management, Res., 4; 105-119.

Hee-jung Park & Bhargava, D.S., (1992), Paper on Impacts of solid waste, Punlished in *Indian Journal of Environmental Protection.*

Hyrayama, 1977. "Systems and operation of solid waste incinerator". Japan Environmental Sanitation Center, Tokyo.

JICA, 1983 "Seoul City incinerator basic design report, Seoul City," Korea.

Parivesh, Newsletter, published by: *ENVIS Center*, CPCB, Parivesh Bhawan, East Arjun Nagar, Delhi-32.

Stewart, R.D.(1975) "The effects of CO on humans", *Annual Review of Pharmacology for nitrogen oxides*", AP-84, Washington, D.C.

Stoker, S.H. & Seager, S.L. (1976) *Environmental Chemistry*, Scott Foreswman & Co. New York., Vol. 15. 409-423.

United States of Environmental Protection Agency. (1971) *Air quality criteria for Nitrogen oxides*. AP-84, Washington, D.C.

6

ENVIRONMENTAL REPORTING

A Green Initiative

S. Kannan and *A. Padrakali*

The environment, in which we live today, is not the same as several decades before. Human activities through the years have degraded all the vital life supporting systems like air, water and land. During the last three decades various laws and conservation activities have been imposed. 'Green' movements have become popular in Australia, Europe and US. In India realizing the importance of environmental protection, National Environment Policy was framed.

Draft National Environment Policy

The Union Ministry of Environment and Forests released the draft National Environment Policy. The draft National Environment Policy touches upon the key environmental challenges facing the country. The objectives of the environment policy provide the strategic themes for intervention, broad indications of the legislative and institutional development intended, the mechanism for implementation and review.

This paper is a attempt to outline the Environmental reporting as a measure to protect environment by creating environmental consciousness.

Corporate and Environment

Business affects Environment in two ways :

(i) Consume energy and other resources, like sunlight, water oxygen, carbon dioxide, raw materials, human resources, rain, sand and clay.

(ii) Create wastes that must be disposed of preservation of Environment.

Manufacturing and other trading companies that emit harmful chemicals and other particles, affect natural environment. As a result of pollution, it was discovered that the ozone layer over the south pole had started disappearing at a very fast pace giving space to the harmful ultraviolet rays of the sun to travel all the way down on earth. The challenge of environmental degradation-cum-environmental pollution and change of sustainable development must be considered as the two facets of the same challenge. Hence, it is necessary for each and every company to take proper care of the environment around it. The companies in today's scenario of increasing threats to the environment must, along with taking care of their profits, take important steps for the protection of the environment in and around them.

ENVIRONMENTAL REPORTING SHALL BE A MEASURE TO CREATE ENVIRONMENTAL CONSCIOUSNESS AMONG CORPORATES

Content of Corporate Environmental Reporting

A sound corporate Environmental reporting shall have the environment policy of the company, details regarding the key executive in-charge of environment division, the goals and targets on environment in concrete terms, legal compliance of environmental issues, Research and development of company on environment, Environmental management systems and programmes and initiatives for environmental protection.

Environmental Reporting

Accounting and reporting of environmental matters have been

rapidly emerging as an important admissions of corporate reporting. The object of providing information by segment is to provide users of financial statements with information of the relative size, profit contribution and growth trend of different industries and different geographic areas in which a diversified enterprise operates to enable them to make more informed judgements about the enterprise as a whole.

Attempts in Environmental Reporting

Certified Research Reporting of the Chartered Association of the Certified Account of UK had made an early attempt of developing a system of external environment reporting based on the findings of the *United Nations Inter-Governmental Working Group of Export on International Standards of Accounting and Reporting.*

Types of Environmental Issues

The type of environmental issues that are pertinent to the enterprise and its industry are :

1. The formal policies and programmes that have been adopted by the enterprise with respect to an environmental protection measures.
   For example, industries like the Asian Paints India. It carry out programmes for the safety of the ecology by taking care that the harmful chemicals do not affect the environment adversely.
2. In case where no such policies or programmes exist, such a fact must be stated in the audit report.
3. The improvements in key areas that have been made since the introduction of the policy or over the past five years, which is shorter.
4. The environmental emission targets that the enterprise has set for itself and how the enterprise is performing related to that target.
5. The extent to which environmental protection measures

have been undertaken due to governmental legislation and due to which governmental requirements are being achieved.

6. Any material proceedings under environmental laws management should disclose to know and potentially significant environment problems until it can objectively conclude the problem is not likely to occur, and if it does, that the effect is not likely to be material.
7. The financial or operational effect of environmental protection measures on the capital expenditure and earnings of the enterprise for the current period and any specific impact on future periods.
8. When material, the actual amount changed to operations in the current period together with a description of the environmental measures to which they relate.
   The amount may be sub-divided in the following general ledger accounts : (i) Liquid effluent treatment, (ii) Waste gas and air treatment, (iii) Solid waste treatment, (iv) Analysis, control and compliance, (v) Remediation, (vi) Recycling, (vii) Others.
9. In places where it is not possible to segregate the amount that relates to environmental protection measures, this fact could be stated.

Framework for Environmental Reporting

A recent study of the Institute of the Chartered Accountants in the UK and Wales, recommended a framework for environmental reporting which includes the following elements (a) Details of the corporation's environmental policies not in the essence of an immediately attainable goal but more as ideal to which the corporation will continuously strive; (b) A statement of concrete objectives preferable stated in specific auditable terms; (c) Narrative disclosure of the companies activities, which add core business issues rather than peripheral matters; (d) Quantitative disclosure of the corporations environmental impact. The data should be both technical and financial.

Survey on Indian Industries Reports Regarding Environment Protection

Indian position is surveyed with reference UNEP, Industry and environmental programme, which has classified the corporate environmental reporting in the five parts :

1. Green glossaries, new letters, video's in short statement in annual report.
2. Special initial environmental report often linked to a first formal policy statement.
3. Environmental reporting linked to the environmental management system, but more text than figures.
4. Provision for full TRI style performance data on annual basis corporate and site reports. The environmental report is referenced to in the annual report.
5. Sustainable development reporting. Its aim is no net loss of carrying capacity, linking of a environmental, social, economic aspects of corporate performance supported by indicators of sustainability, integration of full cost accounting in financial reporting.

Charter of Ministry of Environment and Forests

The Ministry of Environment and Forests formulated a charter on "Corporate Responsibility for Environmental Protection," a mutually-agreed document between the Government and industrial houses, incorporating voluntary initiatives by 17 identified categories of industries to ensure total compliance with pollution control norms. The charter is the outcome of a four-month-long discussion among the Central and the State Governments, pollution control boards and 17 major polluting industries. This was the first private-public partnership of its kind without any compromise by the Government.

Table 6.1 : Environmental Reporting of Companies in India

Sl.No.	*Name of the company*	*Years*	*Environmental Disclosure*
1.	Asian Paints (India) Ltd.	1998-99	Ecology and Safety
2.	Choliride Industries Ltd.	1998-99	Environment, Safety and Pollution
3.	Colour Chemical Ltd.	1998-99	Safety and Ecology
4.	Goodlass Nerolac Paints Ltd.	1998-99	Pollution
5.	Maruti Udyog Ltd.	1998-99	Environment
6.	Steel Authority of India Ltd.	1998-99	Environment Management
7.	Bajaj Temp Ltd.	1998-99	Environmental Audit
8.	IOCL	1998-99	Safety, cons. of Energy
9.	BPGL	1998-99	Safety, cons. of Energy
10.	The Indian Cements Ltd.	1998-99	Safety, cons. of Energy
11.	The TATA Steel	1999-2000	Safety, cons. of Energy
12.	Henkez Spic India Ltd.	2000	Safety, cons. of Energy
13.	Hindustan Zinc Ltd.	2000-2001	Safety, cons. of Energy
14.	Clariant Ltd.	2000-2001	Safety, cons. of Energy
15.	Torrent Pharm Ltd.	2000-2001	Safety, cons. of Energy
16.	Toto Pumps Ltd.	2000-2001	Safety, cons. of Energy
17.	India Morters Ltd.	1998-99	Safety, cons. of Energy
18.	Nirma	1998-99	Safety, cons. of Energy

Source : Dr. L.N. Loli, *The Mangement Accountant*, July 2002, P. No. 509).

Conclusion

Analyzing the environmental disclosures, it is clear that environmental reporting is at its infancy in India. However, even this type of brief statement in the directors report is also absent in most of the annual reports. In its environment statement the concerned industry is required to provide information on (a) Water and raw material consumption, (b) Pollution generated, (c) Nature of hazardous and solid wastes and disposal practices, (d) Impact of pollution control measures on conservation of natural resources. Considering the poor level of environmental disclosure in the corporate environment reports, the possibility of constraints of inclusion of the environmental statement in director's report should be judged. Thus the Environmental Reporting will fulfil the social responsibility of corporates.

REFERENCES

The Hindu, March 13, 2003.
Business India, Heart of Darkness February 8, 1998, pp. 6-68.
Dr. L.N. Loli, *The Management Accountant*, July 2002, p. 509.
Puthia Asiriyan, June 2003.
The Hindu, May 2003.

7

THE ROLE OF MASS MEDIA IN ENVIRONMENTAL DEVELOPMENT

S. Srinivasa Raghavan

> *"The key.. will be a new public awareness of how serious is the threat to the global environment."*

Under the new world order, based on freedom, democracy and free market, and which may change into some kind of centralization detriment to man's freedom, spiritual, political, cultural and economic environment, unless a spiritual and cultural breakthrough which copes with such technological industrial progress takes place. *Such aspired breakthrough is the role of mass media that put before the public opinion the actual problems of the world which jeopardize man.*

Both developing and developed countries are facing serious environmental problems. While some of the problems might be alike due to their global nature (e.g. global warming, depletion of the ozone layer), others are specifically local ones. How do people out of different cultures perceive these problems? How do they react to them?

Information ensures that environment is in the service of all. Such information is called environmental information which seeks to achieve the goals of protecting the elements of environment, by affirming the positive aspects of our attitude towards environment and adjusting the negative attitudes, through supplying the data and facts related to environment, with a view to changing man's bad conduct which harms the environment.

There is one question particularly, which we would like to emphasize: most people have a very scare and uneven knowledge about demographic problems, because our society has not developed a solid demographic culture able to communicate and support knowledge in this field. This is true, even though the attitudes, strategies and choices connected to these facts touch the lives of all of us. In addition, the sources of information available to the general public are generic articles and programs distributed by the media. Only scholars and students have access to more articulate information.

There are two types of problems related to this state of general knowledge. The first concerns the communication of information which is complete and correct, while the other concerns the form in which the information is communicated. Another way of saying this is that we need to consider both the contents and the language of an article or a broadcast. A third question concerns how the information is received and understood. We want to understand what effects the information transmitted by the media has on those who read it or hear it. We want to see how individual values are formed, evolved and developed. We want to know-how these values operate in the family and in the social context. Finally, we want to understand how the process of compromise between the individual and society is affected by the incessant stimuli of communication systems.

The role of media in creating environmental awareness with the underlying concept of media is very broad one. The different media sources considered can be categorized as follows:

(a) Mass media (e.g. radio, television, newspaper and magazines)
(b) Institutional media (e.g. school, government officials or village leaders, District Information Centres)
(c) Traditional media (e.g. family members, other relatives, friends and neighbours, NGO work)

First, we shall analyze the promotion of environmental awareness by the modern mass media. Finally, we shall see a plan of action—some positive action plan.

It is still to be confirmed that the role of mass media is one of the most important factors underlying the knowledge of environmental problemmes : This can only be true on the conditions that first, mass media are accessible to large proportions of the population, second, are spending some time on environmental issues and third, people are interested in information on ecological issues provided by the media so that they view or listen to the corresponding programmes as well as read newspaper articles or other written publications dealing with environmental issues.

The idea of environmentalism

The idea of environmentalism is a crucial, and yet, a deeply troubling one laden with political overtones, cultural diversity, differing social values all in collision to enhance, or in collision to prevent, the true understanding of the problems enbedded in it. The world is yet to accept the fragile and delicate nature of our environment; the urgency to resolve its problems is not commonly shared by all. The problem of news media's outlook on the issue is one of commercial news values and the very tenet of objectivity. Environmental beat is a usual suspect in the newsroom with new managers, with some exceptions, perceiving it as a radical Green Peace idea that forces them into advocacy route. Process takes time to explain and evoke public interest. Many news people simply stay away from it. So goes the newsroom argument. To compound the problem the environment as a news beat story does not seem to have a clear definition, a clear focus.

Learning Issue Awareness

Students frequently report mass-media, especially television, as their major source of information on environmental issues. The exception to this finding tends to be when schools have established formal course experiences dealing with issues.

Studies regarding sources of information on environmental issues for adults indicate the same pattern; mass media, especially television, is the major sources for them. Magazines and newspapers are also cited as mojor sources of information by adults. Strategies that have been found effective for increasing awareness of issues by use of mass media include case studies, trend analysis, and issue analysis.

Role of Mass Media in Creating Environmental Awareness

The mass-media considered here trying to explain environmental awareness are newspapers, radio and television due to the fact that these are the most widely and frequently used modern mass media in our country. First, I will deal with the supply side, the coverage of environmental issues in the media, second, we will take a look at the side of demand, considering the aspect if people are interested in the offered information on environment and use it.

Print Media as Awareness Generator and Information Provider

Since independence, the number of Indian newspapers has multiplied several folds. Today, Indian newspapers are brought out in 93 languages and dialects. In a recent household survey, 57 per cent of the interviewed persons stated to read the newspaper for 30 to 60 minutes per day. With an average of about 50 minutes the daily amount of time spent on reading the newspaper, "the backbone of news" in India, is considerably high.

While the figures mentioned above underline the importance of newspaper in spreading information, its limits should be mentioned, too : illiteracy in India is around 51per cent. In addition to that, 93 per cent of the several million copies are sold in the large towns with population over 100,000. This accounts only for 10 per cent of the population.

How extensively are environmental issues covered in the daily newspapers? After *Times of India* has abandoned its weekly column, none of India's daily newspapers has a regular space reserved for environmental news so that these are only addressed as government, ministry and legal news as well as in contexts of science or health. The focus is mainly on local and national issues like air, water and noise pollution as well as deforestation. In contrast to the global environmental problems are only mentioned very rarely, e.g. if they are the topic of international conferences.

In general, the demand of journalists when writing on evironment is neither an educational or motivational approach nor offering scientific explanation or background information [exceptions are environmental issues that are on the top of the political agenda, e.g. CNG (compressed natural gas)], but it is to inform their readers on the latest, trendes in political news related to environment. A slight tendency towards a motivational coverage of environmental issues can be observed in the regional language newspapers, e.g. when enlightening farmers on the negative consequences of the use of pesticides.

Encouraging Role of Print Media

Roughly it can be said that India's most important news paper *Times of India, The Hindu, Hindustan Times or Indian Express* have an article on environmental issues at least every second day with *Times of India* doing the best job in terms of frequency. The low level of coverage on environment can be explained by different reasons. First, the fairly new field of environmental journalism which starting point was only in 1994 after the Bhopal catastrophe is not considered as an attractive one by journalists, the majority of the journalists covering environment has been asked to do so. Of course, there are several environmental journalists who have chosen their subject out of true commitment.

Second, environment is competing with all the other issues covered in the newspaper whereby the most severe section criterion is the news worthiness of an issue. The interest of the reader as far as news on environment is concerned is supposed to be still fairly low by the editors. Nevertheless in the same survey all of the interviewed editors and journalists felt that there has been an increase in the number of articles on environment both due to a higher awareness and interest of their readers (this is especially true for the younger generation) and the visible deterioration of the state of nature.

Although its number of readers is only around 16,000, India's only environmental magazine *Down to Earth* has to be mentioned when writing on environment and media, because it fills an important lack in Indian media and their coverage of environment. Started in 1992 this magazine is a kind of chronology of Indian environmental history. The outstanding aspect of is approach is not only the coverage of a broad variety of environment related topics (ranking from policy to science, from local to global level) and their scientific background, but also the belief that addressing environmental issues in a developing country like India is not a matter of luxury, but of pure livelihood.

Role of Akashavani —'The All India Radio'

Radio is cheap, most easily accessible and radio signals cover almost the whole county. However, on average there are only 4.4 radio/transistors sets per 100 persons. 80 per cent are in urban homes, only about six million sets left with 525 million rural population. Furthermore the frequency of listening to the radio is relatively low. In a recent survey only 35per cent of the interviewes declared to listen to the radio regularly, most of them ranking between half an hour and two hours per day.

In addition to the relatively small number of people listening to the radio the role of radio in creating environmental awareness can not be considered an important one due to the fact that there are only very few regular environmental programmes, although there is an instruction by Supreme Court for all media that programmes on environment should be broadcast.

Motivated by the Ministry of Environment and Forests Delhi FM is broadcasting two weekly programmes on environment, "Kinare- Kinare and "Ao Dilli Savaren". Otherwise, pollution of environment is only unsystematically tackled in programmes designed for special target groups, e.g. rural population, industrial workers or in city programmes, in woman's or children's programmes.

The issues mentioned and the way they are dealt are naturally adjusted to the target group and the background of the programme:

science programmes focus on scientific explanations and new technical development, children's programmes have a more educational and motivational apxroach, city programmes are addressing local problems like air, water and noise pollution and so on. On the national level of broadcasting news on environment are very scarce, if news on environment are broadcast this is most often at the regional level.

In sum, radio is well below its potential in creating environmental awareness and over the last few years no increase in broadcasts and environment can be observed. Nevertheless, there has been at least one progressive approach in the use of radio in environmental education that should be mentioned. In 1998, All India Radio, the Indian National Radio Network, addressed environmental issues such as water, air and noise pollution, deforestation, solid waste disposal, organic farming and other topics in a 52-episode entertainment-education radio serial "Yeh Kahan Aa Gaye Hum" ("Where have we arrived?"), a story of rural background in which a factory settles near a s mall village.

The Positive Role

The entertainment-education communication strategy purposely designs and implements a media massage to both entertain and education in order to increase audience members' knowledge, create favourable attitudes and change behaviour. "Yeh Kahan Aa Gaye Hum" was produced under the leadership of Mrs. Usha Bhasin with the assistance of the Central Pollution Control Board of India. This serial was broadcast weekly with repeated versions from June to December via 31 radio stations covering seven Indian states in the densely populated Hindi-speaking areas of northern India.

Due to the wide preprogramme publicity via radio, television, press and NGOs working in the broadcast area it was able to attract a listenership of around 100,000 persons. While the serial itself addressed environmental issues in an entertaining manner, trying to contact the listeners in their everyday life context, the epilogues usually delivered by a credible media celebrity advertised the educational message. The approach was an interactive one.

A competitive spirit was fostered amongless by awarding prices for the quality of provided feedback, outstanding community work by listeners in the realm of environmental conservation was also recognized. Impact studies accompanying the broadcasting process showed that the radio serial provoked not only past-broadcast discussion, but also the founding of clubs in which members listened to the serial collectively and started campaigns to save the environment.

"Yeh Kahan Aa Gaye Hum" is an excellent example that the use of entertainment communication strategies in the field of environment can be equally successful as in social ones in which it has been practiced before. In my opinion, this approach especially, but not only via the medium radio that is able to reach the less educated and rural parts of the population, is an extremely promising one so that its chances should not be missed in the future.

Television and Electronic media — DD and Other Channels

Due to the high level of illiteracy in India, electronic media are in a key position. Furthermore, inquiries have shown that both radio and television are perceived as authoritative and friendly media by vast percentages of the population. Since India's 45 million television sets cover 86.5 per cent of the country's population, television is one of the most powerful effective means of propaganda and persuasion.

Over 10 million Indian homes have cable and satellite connections. Television has become a new status symbol even in a villages. In ruraly areas people have at least sometimes access to television, e.g. in small restaurants and long route busses. Probably, the rapid growth rate of television sets (3 million annually) will even increase the importance of television in the future.

A further advantage of television as an educational media is that it is helpful in teaching practical work: A televised presentation can be as clear as a face-to-face demonstration. With an average amount of 138 minutes spent on watching television per day and more than 50 per cent out of all the survey respondents watching television two to three hours per day, television dominates newspaper and radio not only in the frequency of usage, but also in the amount of time spent on it.

Considering these promising prerequisites for the use of television in environmental education it is good news that — compared to newspapers and radio - television does a good job in covering environmental issues. Discovery Channel, National Geographic Channel and Animal Planet Channel are broadcasting exclusively on endangered species, wild and sea life. The national channels are regulated by law to offer environmental programmes. Doordashan the most important channel of dissemination that reaches all over the country has three regular programmes on environment: each of the weekly broadcasts of scientifically profound "Earth Matters" focuses on one special environmental problem.

Targeted at educated people "Earth Matters" offers information on the sources, explains the consequences of the problem considered and gives suggestions for solution by changing individual behaviour.

Weekly "Terraquiz", India's first environmental quiz show on television, is a competition of the top ranking schools from Green Olympiad and is sent between Hindi and English news. "Nature Plus" is even broadcast daily, most often however at midnight. The broadcasting of a forth weekly programme of 30 minutes named "Earth" in form of a magazine that subsidizes successful initiatives of individuals is planned by the Ministry of Environment and Forest in cooperation with Doordarshan.

In addition to that BBC's "Earth Report" offers exclusively information on environment and with the daily broadcast of "The New Adventures of Captain Planet" on Cartoon Network there is at least one programme on environmental issues especially designed for children.

As for radio it is true for television that even if there is no consistent programming on environment ecological issues are tackled in the news as well as in children's, health, science, rural, agricultural and educational programmes.

In contrast to radio and newspaper, politicians have recognized televisions' potential in creating environmental awareness so that the Ministry of Environment and Forests grants special money for the production of films on environment. Although adds on television are highly cost intensive the Ministry of Environment and Forests has proposed to Discovery Channel to produce 50 seconds of informational adds on nature and pollution that should be broadcast four times per day and changed monthly.

The "Sanctuary Asia Magazine" with its online service also serves the interest of those who would like to promote the awareness. Sanctuary Asia, India's leading wildlife, conservation and environment magazine, was started by Bittu Sahgal in 1981 to raise awareness among Indians of their disappearing natural heritage. The overwhelming response to the magazine led to the birth of Sanctuary Cub, a children's nature magazine, in 1984 and to *The Ecologist Asia* (Indian edition of *The Ecologist*, UK) a journal dedicated to the issues of the environment, development and human rights, in 1993.

In the 1980s, Sanctuary Films produced two wildlife/ conservation serials aired on Doordarshan, India's national television network. The first, Project Tiger, was a documentary while the other, Rakshak, was a narrative serial for children. The films were shot on 16 mm and the Sanctuary team visited virtually every wildlife haven in India.

The Demand for Environment Information

Having discussed the medias' supply of information on environment, this paragraph will focus on demand side we use the same survey for the purpose. To what extent do people perceive environmental information provided by the media? What kind of medium is most extensively used?

A vast majority of 80 per cent of the respondents claimed that they were informing themselves regularly on environmental issues. Out of these 80 per cent, 93.8 per cent used television, 82.7 per cent used newspapers while only 35.8 per cent mentioned radio programmes as a source of information.

These figures, once more underline that television and

newspaper play a more important role in environmental awareness creation than radio does, both out of the perspective of supply and demand on information on environment. Seventy per cent of the interviewes stated the amount of information on environment to be exactly right, 24 per cent felt that it was too low.

To sum up, we have to ask the following questions:

- What role do Indian media play in creating environmental awareness?
- What are their weaknesses, what their chances?
- What could be done to enhance their effectiveness in raising the level of consciousness of ecological problems?

Role of Indian media play in creating environmental awareness

A common weak point shared by radio, TV and newspaper in India is that there are no journalists specialized (only) on environmental issues, that journalism on environment is neither a part of the education as a journalist nor is there a possibility of further training in this field. Consequently, the frequency and quality of coverage on environment depends to a high degree on the personal interest and commitment of the journalists and editors. Further typical limitations of modern mass media in India are the language, the educational and the urban-rural biases that make it necessary to design special programmes for all the different target groups if the aim is to reach a vast part of the population.

A general problem in using media for education is the lack of feedback and control of the audience or readership so that it is difficult to assess the impact of educational broadcasting. A support system that is needed for an educational message to deliver it to its target group can be offered only in few cases, the same is true for a deepening of information gained through the usage of media, e.g. via sharing experiences through discussions.

Interestingly, only television and radio have an educational and motivational approach when dealing with environmental issues, while newspaper editors feel that their job is limited to providing their readership with information.

The active campaign policy of the Ministry of Environment and Forests, the guiding jurisdiction of Supreme Court and the activism of NGO's have been able to promote news on environment more forcefully so that an increase in the coverage of environment over the last few years can be observed for both television and newspaper.

Weaknesses and Chances

A definite obstacle for environment playing an even more prominent role in media is commercialization. Since many editors feel that "you can get no money out of environment", environment is not an agenda. Despite of all the limitations for media use in increasing environmental awareness of the population, media role in that respect is an enormous one.

Enhance Their Effectiveness in Raising the Level of Consciousness of Ecological Problems

The focus of Indian media when covering environment is definitely on local and typical Indian problems like water and air pollution, deforestation and on the campaigns launched by Ministry of Environment and Forests (e.g. against the use of crackers at Deewali or the promotion of bags out of natural materials instead of plastic) as well as government politics while ecological problems at the global level are neglected to a great extent. It is amazing to see that people asked to brainstorm the environmental problems they have heard about mention exactly the issues covered in Indian media as it was revealed by the household survey conducted for this study. This fact gives strong evidence to the power of media in communicating environmental issues. People themselves seem to be convinced of the dominant role of media in creating environmental awareness anyway: In a recent survey more than half of the interviewes (52%) declared that the information provided by media has been most important in making them aware of environmental problems.

Due to the fact that environmental education via media is supposed to play an especially important or even dominant role in creating environmental awareness the focus of this study on the contribution of media to the prevalent level of environmental awareness has been chosen. Nevertheless there is no doubt that there are additional factors influencing the level of environmental awareness.

Level of Environmental Awareness Among People

Although all would agree that environmental pollution is an extremely urgent problem, their confidence in this belief weakened when comparing the priority of environmental pollution with other problems Indian society has to face. Some are unemployment, insufficient education and poverty, crime, corruption and religious or social conflicts as more urgent problems to be tackled than environmental pollution.

It is striking that people do think of the problems mentioned above as interrelated, but do not see the connection between environmental destruction and many of India's urgent problems. This is probably one reason for the fact that priorities are generally not set on environment.

An education policy launched on rising a higher level of environmental awareness should stress addressing the causal relationship between destruction of environment on the one hand and social as well as health problems on the other hand and should point at the fact that especially the poorer, less educated people are exposed to environmental pollution. Making people aware of the relatedness of environmental and social problems could push the issue of environment more towards the top of the agenda.

Why are Media such a Powerful Instrument in Creating Environmental Awareness?

Trying to find out on the knowledge of environmental problems that can be considered to be in a two-way relationship with environmental awareness (both as prerequisite and as consequence of awareness of environmental problems) the results showed that there is still a lot of work to be done in informing people on the major ecological problems.

This is especially true for those on a global level like global warming or depletion of ozone layer. Renewable energies or conservation of biological diversity, aware of problems like air pollution, water pollution and household or industrial and chemical wastes that are not properly collected and recycled or disposed.

Air pollution was considered as the most serious problem by a vast majority of people, followed by water pollution and waste. These are strong evidences for the fact that the efforts in educating people on environment via media have been fruitful

It is not only that the average amount of time spent on watching television and reading the newspaper per day is a tremendous one, but it also seems that the environmental issues tackled in media are realized and reflected by viewers, readers and listeners at least to some extent, while those issues that are more or less neglected by Indian media, literally the global environmental problems, are not at all present in the public's consciousness.

Furthermore, media, in principle, are able to reach a vast percentage of India's complex society. Prerequisite for this is that environmental programmes are especially designed for all major target groups (age, level of education, language, urban and rural population and so on). In this context, the role of radio should not be underestimated, although the number of listeners and the amount of time spent on listening to the radio is obviously lower than the corresponding figures of television and newspaper: Most of India's rural population does not have access to television and newspaper, India's high percentage of illiterate people cannot be addressed to via the medium of newspaper.

Although the coverage on environmental issues has increased over the last few years, the media's supply of information on environment is still quite low and one-sided: the variety of environmental topics tackled should increase, the reporting should not only highlight the environmental issues that are at the top of the political agenda, but the coverage should be a more continual one. To realize these suggestions media have to be sensitized that, in the field of environment, the mere offer of information is not sufficient. Concerning environmental issues the role of media should not exhaust in providing information, but there should be a motivational and educational perspective, too. Only if consequences of environmental pollution are showed very clearly and connected to the individual's behaviour, the information on environment provided by the media will be able to motivate people to behave in a more environmentally friendly way. The approach of entertainment-education is, in my opinion, a very promising one in creating environmental awareness via the usage of media: it should be used more extensively and spread to all the three different kinds of mass media considered, television, newspaper and radio.

Due to the general problems that have to be faced when media are used in an educational approach, e.g. the lack of feedback, interactive communication, control of the audience as well as the fact that media's approach will always stay a theoretical one, enhancing the efforts in creating environmental awareness should not be restricted to media alone, however powerful they may be, as an instrument of environmental awareness creation.

A Positive Action Plan

The role of environmental information is fulfilled through an information plan that depends on most sophisticated means of communications, information systems and methods of storing and transferring data, so as to enlighten public opinion; besides the necessity of the establishment of a mechanism to implement this plan.

The Environmental Information Plan Aims at :

- Simplifying environmental data
- Explaning patterns of conduct that contribute to solving the problems of environment
- Highlighting the positive aspects in local, regional and global environment work

- Spotting the global environment problems particularly those affecting us; Ozone hole high temperature—acid rains (....etc.)
- Focusing on the relation between development and environment and confirming the concept of sustainable development as a basis for achieving economic growth, not destructive growth
- Focusing on the population problem and enlightening people on the dangers of overpopulation

First : Printed Information

The Environment Affairs Agency may issue many bulletins, pamphlets, booklets and posters, as well as the green message; in periodicals to mediamen and pressmen. The plan :

- Newspaper allocates a news coverage on environment.
- Allocates a weekly page on environment, through which it presents environmental competitions in information and painting.
- The State Information Service issues many booklets on environment, natural preserves and tourist sites.

Second: Audio—Information (Broadcast)

Broadcast presents many short programs characterized by information-giving and simplifying the masses' participation, such as: "My Environment is My World" on irrational conduct, dangerous and harmful practices to which man may expose in the street. This programme should allow the masses' participation through several means such as: competitions in photography, or painting or research and short story.

Third: Visual Information (TV-Video-Documentary Films-Cinema)

It may be obvious to say that the mass media in general and

television in particular, have a huge cultural significance, since television is the most popular and ubiquitous popular medium, offering diversity and availability unmatched by the press. The way in which the audiences of mass media and television interpret the world shapes their existence and their participation in society. Television can be considered "the site of convergence that joins the private world of the home with the larger public worlds beyond the front door" TV as a mass media depends on picture and sound, hence, promoting the conveyance of the information message. Such country-wide media presents a group of programmes which form a significant part of the information plan.

A Plan for TV

For instance,

Channel 1 presents a weekly program "you and life". It conducts discussions with citizens on how to preserve the environment and combat pollution.

Channel 2 presents a number of programmes such as:

- " Its One World" : a weekly programme for all categories.
- "Development and Population": a weekly programme for all categories.
- "Watch Out", presented twice daily.
- "Motherhood and Childhood" : weekly programme for all family members.

Channel 3 presents a number of programmes on environment such as:

- "Reportage": a daily program for all categories.
- "The Green Arrow": a weekly programme for all categories.
- "Attention! Family": a weekly family programme.
- "A Song for My Country": a weekly family programme.

The information plan focuses on students' gatherings and

cultural centres as important sites for spreading environmental awareness, through open seminars and conferences.

1. Giving priority to environment issues in the magazines published by the various vocational unions.
2. Resorting to orientated information campaigns under attractive slogans.
3. Paying attention to preparing and producing children programmes aiming at creating new generations who have environmental awareness, through cartoons, puppets, games and stories.
4. Producing documentary films on environment issues and natural preserves and project them in cinemas, cultural centers and seminars.

We will need to modify our approach according to what we want to find out. With regard to this, we surely have to enlarge our theoretical field of vision and the type of analysis we use, in concert with disciplines that study communications and their effects. If instead we want to use media for specific communications questions, we must understand the mechanisms specific to each type of medium, so that we can gather information about the ways information is received and the impact of communications. These mechanisms will include such things as targeting user/consumers, content analysis, and linguistic analysis.

Obviously, it is largely the job of the media to spread information detailed and accurate. Their role is particularly important since the education system has so far neglected to foster interest in and sensitivity to these questions. Two types of intervention are needed: on the one hand we need to raise the consciousness of people working in the various media so that they will give this information the emphasis it deserves. They also have to insure that it reaches all segments of the population.

REFERENCES

A study on Environmental Awareness and environmentally beneficial behaviour in India, April 2002.

Bandhu, Desh and Dyal, R. (Eds.) (1999): *Environmental Education for a Sustainable Future*, Indian Environmental Society, New Delhi.

Das, Nandita (2001): *A study on social and environmental impact of TV and radio programmes*, Centre of Media Studies, New Delhi.

Das, R.C., Baral, J.K., Sahu, N.C., Misra, M.K. (1998): *The Environmental Divide - The Dilemma of Developing Countries.*

National Council of Educational Research and Training (1997): *Fifth Survey of Educational Research 1988-92*, Vol. 1.

Public awareness and the role of media: An Italian survey on the 6 billion day.

8

DESIGNING A MASS MEDIA CAMPAIGN FOR ENVIRONMENTAL DEVELOPMENT

K Tamizh Jyothi, C. Samudhara Rajakumar and *C. Madhavi*

Introduction

Life is a gift and human life is precious gift. A small human brain keeps thinking on innovations and takes the world to a new peak. Mostly, life comes to an end due to accident, natural disaster, chronic and other non-chronic diseases. Among these, most lives are taken by various diseases. Diseases occur to human being even when he/she starts his/her breathe. Over 90 per cent of the diseases occur due to the environmental pollution like air, water, noise etc. We cannot live in pollution less world, except if we live in villages, which is free from air, water and noise pollutions.

In our day-to-day life, due to the technology advancement, we enormously use the electronic equipments for entertainment and to reduce our workload. Examples are Television, Refrigerator, Microwave Oven, Computers etc. Most of the components in the electronic items are made up of non-recyclable plastics. These plastics, when disposal is not friendly to the environment and create air pollution. Also Freon, the gas that present in the refrigerator will create hole in the ozone layer which leads to passing of the cosmic rays to the atmosphere. These cosmic rays are highly toxic. Microwave Oven also produces toxic gas which is harmful to the environment. During image production, in the film industry, huge amounts of toxic waste are produced. These wastes are not only polluting the environment but also ultimately dangerous to the human life. When producing liquor, waste gases are released in the atmosphere. These gases are highly toxic in nature and affect human beings and plants. For example, the tender coconut in Cuddalore SIPCOT industrial estate is reddish in colour due to reaction of chemicals discharged from the chemical industries in that area.

Apart from these, air pollution is the main problem which occurs due to smoke from automobiles. The exhaust carbon from the automobiles pollute the environment and create chronic diseases like asthma. To regulate this, a fuel called Compressed Natural Gas (CNG) has been introduced by the government in New Delhi. In addition to the automobile pollution, some other industries are also polluting both air and water like sugar industries, tanneries, dying industries, etc.

One simple way to overcome this problem is keeping the environment green. Plants produce Oxygen that we need and absorb Carbon dioxide that we release. So it is better to say "Green plants are our friends". That's why many organizations are following the practice of planting trees along the roadsides and other areas. In India, all metros are strictly following and carrying the caption "Keep the City Clean and Green".

In the past, Government has controlled the environmental safety activities whereas now some private organizations are concentrating on these issues, which makes other private companies to follow. To protect the environment, many ways are followed by the corporate, among which Mass Media is an important means of creating awareness. Mass media can easily penetrate into large audience so that any message conveyed to the public is easily understood.

What is Mass Media?

Mass media is the media that carries or conveys the message and information to the target audience. Here, there is no interpersonal communication between the communicator and the audience. Tones (1996) pointed out that mass media campaign have the ability to reach large number of people. He highlighted that there is a wide range of strategies to be considered under the headings of mass media. These include print, electronic, films, leaflets and posters, carrier bags, postage stamps, pens and greeting cards.

This advantage of Mass media can be encashed to use this as one of the best ways of communicating to the target groups about the environmental pollution and problems. To create awareness about the importance of keeping our environment Green, organizations develop drama, skit and telecast in the television or broadcast in the radio. Since sending greeting cards is a growing culture among the educated people, it could be printed with the message like "Save Environment" and sell it in the market. With the help of mass media, the awareness can be created to influence the audience to perceive and think positively towards the message. All mass media tools cannot be applicable for the entire target audience. There are some of the tools which are meant for specific category of people/audience. However, creating Mass Media Campaign (MMC) is a crucial task for any organization. In MMC, seven steps are involved in planning and implementation. The following diagram slows the steps in mass media campaign.

Steps in MMC Development

1. Research the factors related to the environmental problems

The first and foremost step is identifying the factors related to the environment problem. It is based on the nature of the issues and the campaign meant for that problem. For example, if it is a campaign to emphasis not to release the exhaust gas in the atmosphere, then they should identify the cause of this toxic gas, how it is harmful to the environment, public and what are the necessary actions to be taken etc. So identifying the factors is the primary task in any mass media campaign.

2. Identify the target group (Beneficiaries)

Once the factors are identified, then the beneficiaries must be found

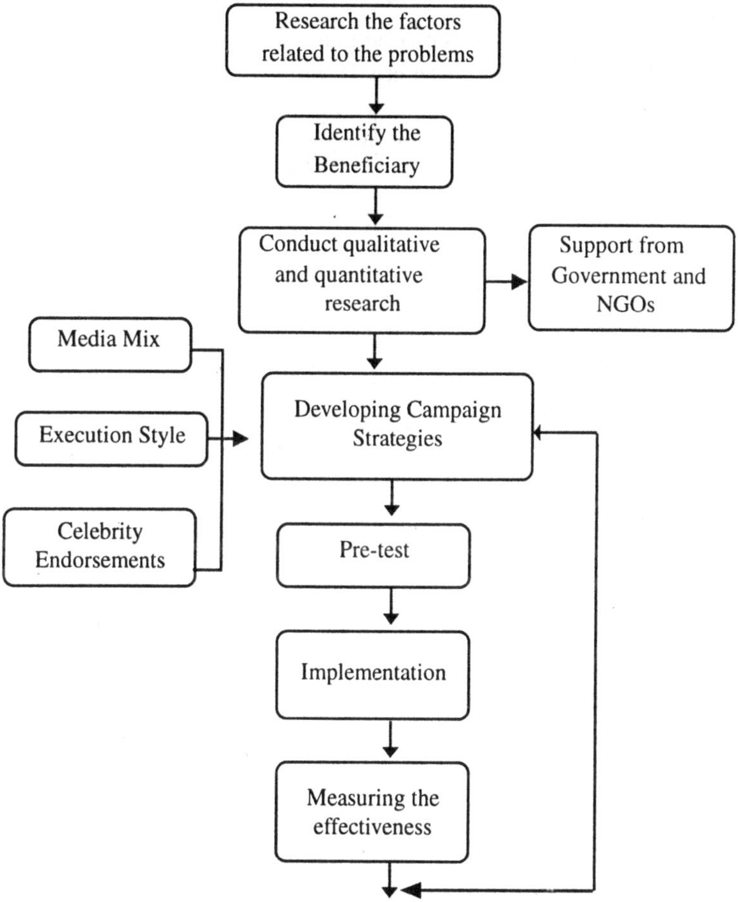

Fig. 8.1 : Steps in Mass Media Campaign for Environmental Development

out, because the campaign is meant for a specific group. For example, in automobile sector, the benefits of using CNG should be informed to the automobile owners and not to the other general public.

3. Conduct qualitative and quantitative research

For collecting secondary data both qualitative and quantitative, necessary data may be collected from government department. For example, awareness about environmental pollution in a particular area, number of industries closed mainly due to pollution problem in an area, level of pollution etc. If we need primary data we can collect by conducting survey. For example, people knowledge hold industries pollute environment and that impact om society.

4. Developing campaign strategies related to the environmental problem

This is the crucial task for any organization. The development of campaign strategies is comprised of three components, viz. Media mix, Execution style and Celebrity indorsements. Each component has its own merits and demerits.

Components in Developing MMC Strategies

Developing MMC Strategies

- Media mix
- Execution style
- Celebrity endorsement

Media Mix

Four types of media are available, namely electronic, print, outdoor and direct mail. Electronic way of sending messages can easily be understood even by the illiterates and the number of audience for which the message targeted can also be high. Here the cost involved is more. If the message is meant for general public, then electronic media can be preferred. In addition to this if the message is targeted to a specific and small group, compact discs (CDs), or Video cassettes can be prepared and sent to them. Print media has the advantage of low cost and high reach. The only drawback of this print media is it can be understood only by the educated persons. Corporate should prepare the printed message in the form of cartoons, drawings, pictures etc. to have attraction and the illiterate can also understand the concept. In cities and towns, one could see the hoardings highlighting the importance of rain water harvesting, planting trees etc which are coming under the Outdoor media Companies should come forward to use this media to create awareness about the various environmental problems and issues.

Execution Style

The message executed contributes to the relevance, originality and impact of the compaign Execution style deals with the kind of appeal the campaign should possess. Organizations can adopt different execution styles like straight forward message, fear appeal, etc.

(i) Straight forward messages: Here the actual message conveys information without any gimmicks, emotions, or special effects.
(ii) Fear appeal: Public is informed about the danger or negative effects of particular message.
(iii) Humorous appeal: By seeing the campaign people enjoy and relish which in turn make them to listen and understand the message.

Other execution styles include emotional appeal, utility, slice of life etc.

Celebrity Endorsements

As the corporate found difficult to gain/attract the attention of the audience, they use celebrity endorsements as a strategy to influence people. Celebrities are famous personalities who are known to the public. Because of their fame, celebrities serve not only to create and maintain attention but also to achieve high recall rates of the messages (Croft *et al.* 1996; Friedman & Friedman, 1979). People love celebrities because the characteristics of the later are liked by them. Studies show that celebrities create positive attitude towards the advertisements or campaign and perceives the message in a positive or highly favourable manner. Social influence theory which includes the source credibility and source attractiveness argues that various characteristics of a perceived communication source may have a beneficial effect on message receptivity (Kelman, 1961). Also the cultural meanings residing in a celebrity go beyond the person and are passed on to the product or campaign (McCracken, 1989). The campaign that uses celebrities will have high impact among the target audience. So, if the organizations want to influence the audience in short period, they could use celebrities in their campaign. Celebrities should be used based on the nature of the campaign and target audience.

5. Pre-Testing

Following the development of MMC strategies, pre-testing must be done. Pretesting refers to testing the campaign with a small getup of target audience or in a particular geograpghical region. Based on the feedback, any alteration on the campaign, can be carried out prior to finalizing the campaign strategy. In case of planting trees, the format of the campaign can be circulated among similar organizations or with social welfare clubs to seek their views for improving.

6. Implementation

The pre-tested campaign strategy, after modifying the pitfalls, is ready for implementation. While doing so, it is suggested to start implementing from smaller area and extend to larger area. Also the campaign should be monitored whether it has achieved the company's social objective or not.

7. Measuring the Effectiveness

Any campaign must be evaluated to find out the progress of it. In the same way MMC must also be monitored and evaluated regularly.

A separate team must be involved in evaluating the campaign. Deviations identified must be bring to the notice of chief campaign officer for incorporating suitable remedial action.

Suggestions

So far mass media campaigns for the environmental development are organized by the government alone. Private organizations are not showing enough interest in these types of activities because the expensive is at the higher end. Since social work is other part of the organizations objective, giving mass media campaigns alone will require huge investment as the advertising agencies extract more amounts which is affordable by few big organization. Obviously, firm show less interest in these kinds of activities. One way of overcoming this hurdle is both advertisement agencies and media owners need to come forward to offer concessions to the organizations which are imparting MMC related to the environmental safety.

In order to motivate the corporate, government should offer Income Tax exemption to firms that give MMC for environmental friendly activities. This will attract many organizations to enter into environmental safety campaign through mass media. Government should assist in all aspects of the programme to the corporate by asking the local government or organization to support with the corporate in bringing the campaign a success.

In European countries, the private organizations are doing the social activities particularly in environment safety. As in other countries, like India also these programmes should be extended to the private organization. Also, the organizations should allocate some percentage of amounts towards the social activities in their annual budget. This will bring our country a "Green India" in near future.

Conclusion

Tons of industrial effluent and sludge generated everyday is inflicting permanent damage to our eco-system. It is equivalent to poisoning ourselves and the future generation. It is high time we act upon immediately and contain the damage. Organizations should come forward to launch innovative mass media campaigns for environmental development. Let us create a better and cleaner environment for the next generation.

REFERENCES

Croft, Robin, Dean *et al.* (1996), *"Word-of-Mouth Communication: Breath of Life or Kiss of Death?,"* In: The Proceedings of the Marketing Education Group Conference, Glasgow: The Department of Marketing, University of Strathclyde.

Eyiah, Joe Kingsley, (2004), "The role of mass media in community development", *Brookview academy*, Toranto- Canada.

Fox, K., and Kolter, P. (1980). The Marketing of Social Causes: The Years. *Journal of Marketing*; 44, 25

Friedman, Hershey H. and Linda Friedman (1979), "Endorser Effectiveness by product Type," *Journal of Advertising Research*, 19 (5), 63-71.

Kandice L. Salomone (1992), *News Content and Public Perceptions of Environmental Risk: Does Technical Information Matter after All?*

McCracken, Grant (1989), "Who is the Celebrity Endorser? Cultural Foundation of the Endorsement Process," *Journal of Consumer Research*, 16 (December), 310-21.

Peter M. Sandman, David B. Sachsman, *et al.* (1987), *Environmental Risk and the Press*, 6-98.

www.acpr.gov.au

9

ISO–14000 – AN EMERGENT NEED FOR ENVIRONMENT MANAGEMENT SYSTEM

K. Kavitha and *A.G. Murugesan*

Introduction

Our earth is polluted through water, land and air due to increasing urbanization and industrialization. Polluted water means that the quality has been degraded by the addition of substances such as chemical effluents, metal residues, oil and sewage. Land pollution is the collective term of variety of process in which terrestrial habitats degraded as a result of harmful substances being released into environment. Air pollution involves the release of substances harmful to living organisms into the atmosphere. So, it is our urgent need to manage and protect our environment. It can be built by using two principles i.e., the balancing principle and precautionary principle (Natarajan, 1999).

Sustainable development on a global scale is our ultimate goal and environmental protection is an integral part of Sustainable Development as declared by Earth summit. International chamber of commerce had recognized environment management as a key to Sustainable Development and to establish policies, programmes and programmes for conducting operations in an environmentally sound manner.

ISO-14000

A systematic approach with emphasis, on continual improvement is essential for achieving sound environmental performance. With advent of ISO-9000 standards during 1987, the ISO-14000 standards hold out a promise to revolutionize environmental protection i.e., it provides all the requirements and recommendations for any organization to develop and implement a cost-effective eco-management system. In other words, EMS helps the organization to manage its activities, products and services in such a way to minimize or eliminate its adverse effects on environment.

Need for ISO-14000

Governmental organizations, world over, have conceded that growing focus should be pollution prevention rather than pollution control. Waste minimization should address environmental issues right from the choice of technologies choosing a low polluting Technology in the manufacture of a commodity. The statutory authorities do not have the competence to dictate to industry how processes should be designed (Saravanan, 2001). Thus the only option left to us for improvement in the environment scenario is to stimulate voluntary action on the part of industry. ISO-14000 series of standards builds) many voluntary programmes adopted by various organizations and provides the framework for integrating strategic environmental management into company operation. Implementations will be verified through third party, ISO-14000 is far apart and above any existing environmental programmes.

IS0-14000 Standards

ISO-14000 is an evolving series of generic standards being developed by ISRO to provide organization with the structure for managing environmental impacts. ISO-14000 Series refers to the family of international standards published by International Organization for Standardization (ISO) in 1996. This specifies the required elements of an EMS. ISO was founded in 1947 (Headquarters in Geneva, Switzerland). ISO-14000 standards are developed by fifty-five members countries with 155 members and seventeen liaison. It has two basic types of documents i.e., Specification standards and Guidance standards form the ISO-14000 series (Akila *et al.*, 1999).

Quality Management Standards (QMS)

Quality Management Standards (QMS) concentrates only on customer's satisfaction and quality can be defined by technical specification, so, customer's satisfaction is quite easier.

Environmental Management Standards (EMS)

It is the part of the enterpriser's overall management system which includes organizational structures, functional charts, planning activities, responsibilities, accountability, practices, procedures, work instructions and resources for developing, implementing, achieving, documenting, reviewing and auditing the environmental policy.

An organization can design, develop and implement their EMS as per the specification of recognized standards. Till June 15, 1994, there was only one international standard i.e., British Standard (BS 7759) developed by British Standard Institution. The Indian Standard on environmental management system issued specification on environment [speficication IS -3967] was issued in December 1993 by the Bureau of Indian Standards. This was based on December, 1993 400220 draft version of 35775 1992. Then ISO-14000 was developed in 1996. This includes a wide range of environmental disciplines indulging EMS, EA, EL, LCA and QMS.

About seventeen different standards are being developed by the ISO ITC 207 (TC-Technical Committee). The BS and ISO international standards are similar and vary from each other only in documentation. EMS based on BS can be easily changed to an EMS based on ISO-14000. Among the seventeen standards,

ISO-14001 is a specification standard and all the others are guidance standards. These standards are essential for developing, implementing, maintaining and reviewing organization environmental policy.

General Elements of ISO-14000

The elements, which are considered as a part of company's environmental management system, were taken in account. They are management commitment, business plans, document control, use of documentation, corrective action, preventive measures, training, continuous improvement, supplier relations, communication with employees and internal audit. The figure 1 defines about the training programme, how it can be held in any organization or society for employees or to the general public.

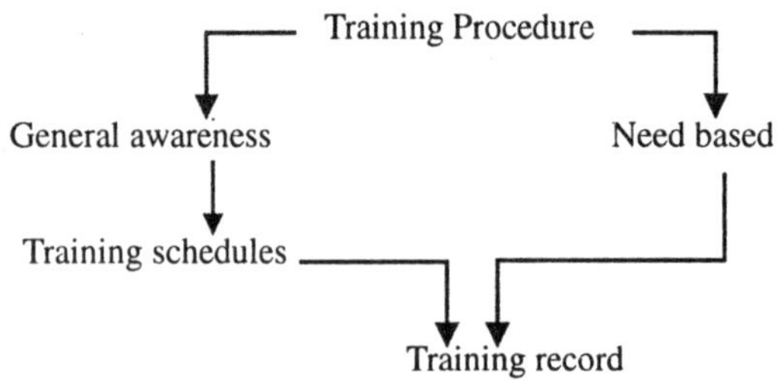

Fig. 9.1 : Training programme held in any organization

EMS

Environment Management System is the facet of an organization overall management structure that addresses the immediate and long-term impact of a company's products, services and processes on the environment. The Corporate EMS performance will give about outline of the role of ISO-14000 (Fig. 9.2).

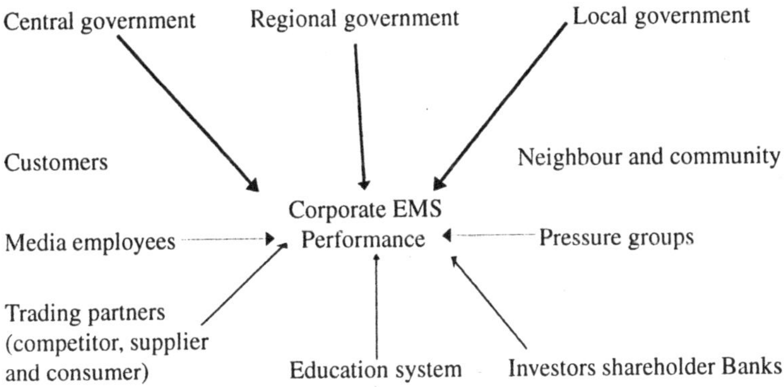

Fig. 9.2 : Performance of a corporate EMS performance

The EMS should satisfy all the above. The scope of application and operation of EMS is wider than QMS. EMS is based on diverse interest.

How EMS can progress?

Progress of EMS in an institution is explained in detail in the Fig.9.3. It was very clear that the EMS can be effectively carried out by the various assessment procedures. Here, first EMS auditing in the external audit which is done by the specific group of auditors initially and periodically it was continued by the internal auditors. The internal audit was documented periodically and kept for verification. Yearly once, the external auditors will check the norms. If it is not within the norms the ISO certification may be cancelled.

ISO-Series

From Table 1, ISO-14000 and ISO-14001 are the two documents, which are considered as the main pillars of ISO-14000 Standards. Next the ISO-14000, ISO-14011 and ISO-14012 are concerned with environmental audit, which is a valuable instrument to verify and improve environmental performance. ISO-14020, ISO-14021, ISO-14022, ISO-14023 and ISO-14024 deals with the labelling

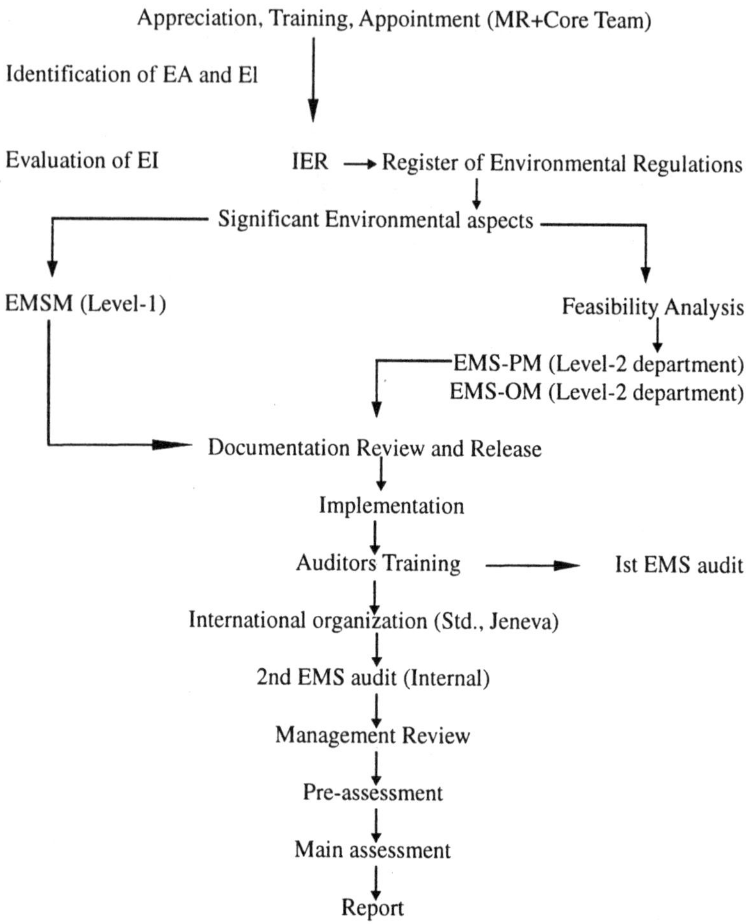

Fig. 9.3 : Progress of an EMS in an institution

standards, which are being developed in order to avoid confusions that may arise from uncoordinated environmental labelling and artificial barriers, if any, to trade etc. Environmental performance evaluation (ISO-14031) will help organization to define and continuously monitor specific aspects of its environmental performance primarily for its own use in planning. Lifecycle assessment standards (ISO-14040, ISO-14041, ISO-14042 and ISO-14043) will deal with each stage of environmental lifecycle of a product from initial design to bill disposition (ISO-14000, 1996).

Table 9.1 : The various documents of ISO-14000 and their role

Document No.	*Document Title*
ISO-14000	EMS-General guidelines on principles system and supporting techniques (Annexure-1)
ISO-14001	EMS-Specification with for use (Annexure-2)
ISO-14000	EA-General principles of environmental auditing
ISO-14011	EA-Audit procedure Part! Auditing of EMS
ISO-14012	EA-Qualification criteria for environmental auditors.
ISO-14020	EL-Basic principles of environmental labelling
ISO-14021	EL-Self-declaration, environmental claims, terms and definition
ISO -14022	EL-Symbols
ISO-14023	EL-Testing and verification methodologies
ISO-14024	EL-Practitioner programmes, guiding principles, practices and certification
ISO-14031	Environmental performance evaluation
ISO-14040	Lifecycle assessment-general principles and practices
ISO-14041	LCA-Goal and definitions, scope and inventory assessment
ISO-14042	LCA-Impact assessment audition
ISO-14043	LCA-Improvement assessment
ISO-14050	Terms and definitions
ISO-14060	Guide for the inclusion of environmental aspects in product standards
ISO-14060	Working for inclusion of environmental aspects in product standards

Source: Akila *et al.*, 1999.

Intangible Benefits of ISO-14000

The benefits obtained by implementing ISO-14000 are as follows :

- Generates economic benefits,
- Create good public/community relations,
- Provides significant competitive advantage,
- Conservation of RMS/energy through recycling/waste/ avoidance efforts,
- Enables to protect human health and the environment,
- Reduces the occurrences of accidents and reduced liability,
- Improved environmental performance is achieved,

- Provides an innovative approach to pollution free technologies with new products, Assurance to customers on the existence of EMS,
- Better management environment impacts,
- Enhances the image of company with higher market share,
- Meeting vendor's certification criteria,
- Insurance at reduced cost is possible,
- Improves relation with Government/NGOs,
- Improves operational efficiency.

Increasing Interest in ISO-14000 Standards

Most of the companies are interested to improve their environment by implementing the ISO-14000 certificate throughout the world. For example, UK-322, Janpan-198, Germany-166, Netherland-119, Denmark-96, Australia-53, USA-34. In India most of the industries had started to get ISO-14000 certification. For example, NLC thermal power plants. In near future ISO-14000 may be made as a precondition for doing business in EU/Asian countries.

Conclusion

Potential for environmental pollution control lies in "preventive" rather than "curative". It is tied to improvement in industrial production processes rather than controlling wastes once they are generated. More emphasis is a "reuse/recycle/recovery". This kind of unique policy is followed in the ISO-14000 i.e., it can act as a preventive tool to control and prevent the pollution load from the various industries.

REFERENCES

Akila, P. Sudalairaja, S., Raja, L.M.V., 1999. ISO-14000 Environmental guidelines. *Proc.Sem.Env.Prob.* pp.46-52.

Environmental Management II. *All India Council. for Management Studies,* Chennai.

Implication of International Standards for quality and environmental management systems, 1997. Survery results.

Ingrid Ritchi and William Hayese. A guide for implementation of the ISO 14000 series on environmental management.

ISO-14000 1996. Environment management standards and implications for exporters to developed markets: private sector development programme, UNDP, New york, 1996.

Natarajan, S. 1999. Vehicular pollution. *Proc.Sem. on Environmental Problems in Tirunelveli,* pp.15-19.

Saravanan, K. 2001. ISO-14000. Environment management system. Eco *friendly, technology for waste minimization in textile industry,* pp.63-68.

Sohrab, 1996. ISO-14000 standards for EMS. NITIE, Udyog, Oct-Dec, 1996.

10

ECO-FRIENDLY ENVIRONMENT AND SUSTAINABLE DEVELOPMENT

S.R. Choodambigai

"We inhabit a single planet but live in several worlds. There is a world of abudance where plenty brings pollution. There is a world of want where deprivation degrades life. *Such a planet cannot survive in harmony with nature and the environment is indeed with itself. We must therefore ensure that the affluence of some is not derived from the poverty of many".*

"The rain, it fallen on the just
As well on the unjust fella;
But it falleth more on the just
Because the unjust stealeth the
Just 's umbrella"

— *Mr. P. V. Narasimha Rao*

Introduction

Sustainability is built into the very concept of income. The concept of sustainability is by no means new in economics, although the word is. Sustainability is implict in J.R. Hick's definition of income as the maximum amount that a person or community could consume over some time period and still be as well-off at the end of the period and still be as well-off at the end of the period as at the beginning. Remaining equally well-off means maintaining capital intact, or maintaining the wealth and population of the community. Growth in Hicksian income is by definition sustainable. Any consumption that is not sustainable cannot be counted as income. Exploiting renewable resources at a profit-maximising sustainable yield is an application of the Hicksian concept of income to resource management. The main idea of Hicksian income is captured in the definition of sustainable development offered by the Brundtland Commission as development that "meet the needs of the present without compromising the ability to future generations to meet their own needs". Thus, sustainable development is about sufficiency as well as efficiency. Sustainability and Sustainable Development are frequently used more or less synonymously but what is being sustained is a level, not a rate of growth, of physical resource-use (T.N. Srinivasan, 1998).

Environment and economic development are inter-dependent and form the essence of the concept of 'Sustainable Development' or 'Sustainable Growth'. The value of sustainability is basic to our economic thinking which guides how much we can consume year after year without eventually impoverishing ourselves. Environment issues have become now a subject of considerable attention within the framework of sustainable development particularly from the point of view of maximization of human welfare relating to a vital question as to how the present generation should utilize scarce valuable natural resources in an optimum manner so that the future generation as a matter of right may get an opportunity for healthy and productive life in harmony with nature. Therefore, the term environment and sustainable development must be treated as complementary.

According to R. Aymong, "the most objective and potentially measurable criterion for sustainable development is the preservation of productivity and full functioning of the natural resource-base" (Raymond, 1992). One may define 'sustainable development' as the "preservation of the production possibilities of an economy to provide the same goods and services including service obtained from the state of nature" (K. Parikh, 1989). The above approach to define 'sustainable development' is highly inadequate as it takes input sustainability as the basic and only requirement for the sustainable development of an economy. The other dimensions of sustainability are not being spelt-out. Also, there is little convergence in the literature on sustainability. For example, according to Little and Mirrlus (1984, 1980) — "this (i.e. sustainability) is more a buzz world - probably derived from the environment lobby—than a genuine concept. It has no merit. Whether a project is sustainable (for over? —just for a long time?) has nothing to (say) whether it is desirable. If sustainability were regarded as a reason for rejecting a project, there would be no mining and no industry" (Little & Mirrlus, 1986, 1990).

The current attempt on Eco-friendly environment and sustainable development is discussed with four sections:

Section 1 : Area of concern of Environment and Sustainable Development;
Section 2 : NEP of 1991 and Environment;
Section 3 : Issues and Recommendations of and Environment and Sustainable Development; and
Section 4 : Functional Relation Between Environment and Sustainable Development.

SECTION 1

Area of Concern of Environment and Sustaintable Development

GDP growth rate reflects the success or failure of the economy as a whole. For providing employment, price stability and material welfare of people high growth rate is essential. It helps in elimination of poverty which in turn reduces environmental stress and pollution caused by poverty. Human population in a country is a source of all economic activities. Unfortunately, it is also a major source of environmental pollution. Availability of adequate quality of water for consumption and various industrial and agricultural uses is to be ensured for sustainability. The Food and Agricultural Organization of the United Nations (FAO) estimates India's rate of deforestation at 0.6 per cent per annum between 1981 and 1990, implying a loss of 3,39,000 hectares over the decade. Each year, 70 million people are added to world population, mostly in developing countries and 15 million square kilometres of forest disappear. This led many people to postulate a simple, displacement model of 'more people' fewer forests (Allen and Barnes, 1985,

Myer, 1987, Ehrlich and Ehrlich, 1990 and Rudel, 1991), but destruction of forests mean loss of watery soils and even life of several animals, birds, etc. So maintenance of appropriate forests coverage ratio in the land for future sustainability of economic development is a must. Human resource development index is a complex one which combines in one number several aspects of human development like economic condition, health care, education, etc. An index showing high magnitude i.e., more than 0.8 is considered to be quite high and less than 0.5 as poor for assessing HRD. Clean air is a must for life and vegetation to survive. But with the rapid development we do find that the ambient air is polluted. The level of air pollution is increasing everyday. Several gases like Carbon-dioxide, Sulphur-dioxide, Sulphur-oxide, Hydrocarbons, etc. pollute the quality of air to a miserable extent. Some of them are responsible for Green House Effects i.e., sustained increase in ambient temperature. Some are responsible for acid rains which are quite harmful for life. The adverse effects of air pollution are to be eliminated for sustainability of the development process. Transport constraints hamper economic development. At the same time transport sector is also an environmental polluter. Both the intensity of transport (transport expenditure / GDP ratio) and there of transportation technology should be consistent with the sustainability.

SECTION 2

New Economic Policy of 1991 and Environment

It is a doubt that the operation of NEP of 1991 had led to resource degradation, loss of biodiversity, land degradation, global climate changes etc. and release of a range of chemical, radioactive and other toxic wastes. Many activities of us had enabled a significant portion of the world to improve the condition of life, but they are showing their adverse effects too on the quality of air, water and land, moving the ecological balance of the planet to a new equilibrium. Whether this new equilibrium is conducive to human activities or whether it will prove adverse effects to the coming generation is a very complex question. But present indications are that there is urgency for taking corrective measures at the global as well as regional levels to prevent environmental degradation or destruction of environment (Dutt. Rudder, 2000).

The outcome of NEP of 1991 like promotion of exports of the products of primary sector, deregulation of industries had direct adverse impacts on the environment in addition to their economic and non-economic indirect effects. For example, under the policy of economic reforms, heavy reliance is being placed on exports as a means to drive the economy forward, but the drive towards this export-led growth strategy had seen sacrificing natural resources to earn foreign exchange which especially has been seen in the fisheries and mining sectors. No doubt, the export of fish and fish preparations increased from Rs. 960 crores in 1990-91 to Rs. 3338 crore in 1995-96, Rs. 4487 crores in 1997-98 and about Rs. 5114 crores in 1999-2000 (Economic Survey, 2000-0l) and that of ores and minerals from Rs. 1497 crores in 1990-91 to about Rs. 2975 crores in 1999-2000, but the commercial ambitions in fisheries have led the Indian authorities to clear a number of 100 per cent EOUs for deep sea fishing by the joint ventures. Increases in mining for raw-material ores' exports has also been taking place in this changed atmosphere. It has been proved that the long-term environmental responsibility suffers when short-run profit maximizing and foreign exchange earnings are the predominant motives for the structural change of the existing economy. During the period 1991-94, 82 companies were given clearance for joint venture marine fisheries using 255 deep sea fishing trawlers (Kothari, 1995). But it is the fact that fish catch is virtually declining in major parts of the entire world, it is increasing in the Indian Ocean proving that the Indian marine economic system and traditional folk are to be disastrous.

The export-oriented agricultural production, i.e. transformation of staple food and subsistence crop lands into cash cropping could damage the environment and the livelihood security of small farmers. Intensive production is highly dependent on the use of fertilizers, pesticides and other artificial inputs for which the poor farmers could not afford. Increases crop productivity through fertilizer, chemical and pesticides caused deterioration of soil health and mismanagement of water had introduced water logging and soil salinity. Further farming on steep slopes caused soil erosion and land degradation. Moreover, ground water depletion is also attributed to target well scheme of NEP 1991. Thus commercialization of agriculture would result in irreversible disequilibrium the environmental scene.

The "delicensing of automobile industry of NEP 1991 has led to a boom in investment in automobile component and plans for producing new cars" (Economic Survey, 1994-95), but it failed to realize that this expansion is occurring at heavy environment cost. No doubt, during the year 1994-95 demand for cars increased by 20 per cent for jeeps 30 per cents, for commercial vehicles 25 per cent, for two wheelers 18 per cent and for three wheelers 41 per cent over the past four years (Kothari, 1995), but such a sharp rise in demand of these commodities definitely adds to the problem of pollution. In absoloute term the sale of cars had increased from 209300 (1993 - 94) to 409624 in 1998-99 and scooters from 833547 in 1993-94 to 1325860 in 1998-99 (Statistics Outline of India, 2000-01).

SECTION 3 (A)

Issues of Sustainable Development

Having understood the effects of NEP and environment it is time to analyse the main issues of sustainable development in a brief manner. The concept of sustainable development raises a number of issues that are briefly mentioned. The major aspects and their related area of concern are briefly noted below:

Revitalising Growth and Sustainable Development

Priority Actions

Accelerating sustainable development; International Policies; Domestic Policies; Integration of Environment and Development in Decision-Making.

Natural Resources

Atmosphere; Ocean and Seas; Fresh water; Land; Biodiversity.

Trans-Boundary Effects

Driving Forces

Values systems and life styles; Population (urban, rural); Socio-economic system; Knowledge.

Production

Energy; Agriculture; Water supply; Industry; Services; Transportation; Forestry; Fisheries; Mining.

Consumption

Level; Resource intensity; Food; Energy; Food; Water; Materials; Other services.

Environment Effects

Depletion of natural resources stocks; Land degradation and pollution; Growing facility of eco-systems; Air, water and marine pollution; Toxic, hazardous and solid waste; Loss of biodiversity; Threats to life support systems.

Human Welfare

Present and future generations

Essential Means Required

New and Additional Financial Resources; Scientific Co-operation and Technology Transfer; International Economy and Related

Domestic Policies; National Capacity Building; Integrating Environment and Development in Decision-Making; Strengthening the Role of Major Groups; International Institutional Arrangements and Regional Organisations; International Legal Instruments and Mechanics; and Information for Decision-Making Source (UNCED, 1992).

In short, to integrate environment and development all national and international policies should move towards revitalizing growth with sustainability (UNCED, 1992) and a balanced approach to development in terms of optimally and sustainability is to be achieved. One way to do this is to hold capital stocks constants from generation to generation. In other words to preserve intergenerational equity, the total amount of capital that is passed on the next generation should not be less than the capital the current generation inherited. Moreover the question of equity between the people of the same generation is also an important issue for sustainable development. The negative links between poverty and the environment should also be broken to have eco-friendly environment and sustainable development (Peace and Warford, 1993).

SECTION 3 (B)

Recommendations for Sustainable Development with Eco-friendly Environment

Development is the process by which people meet their needs and improve their living conditions. Only when such development leads of social, economic and cultural betterment that satisfies the needs and values of all interest groups, not only of the present day but of future also by conserving natural resources and diversity of life, it becomes sustainable.

The goals of sustainable development are not merely the development and well-being of all human beings, but that of all other species also.

Our economy is closely related to the exploitation of nature.

Preservation of nature is this *sine qua non* for economic growth. Ecology which is concerned with the study and conservation of nature and economy which deals with the utilization of natural resources for man's development are interdependent. Proper management of the ecosystem must be an integral part of any programmes for economic development.

In the eco-system, materials are cycled from and through soil, water and air. These operations require energy. This energy originates in the sun. The solar energy energing into eco-system is tapped by green plants through photosynthesis. This energy becomes available to plant when they use the products of photosynthesis. It becomes available to animals when they eat the plants and other animals which eat the plants. The energy is thus realeased and used continuously. Hence, green plants are essential to our planet's eco-system.

Ensuring that human demand on the environment does not exceed the capacity of the environment to support it, bearing in mind the environmental demands of other species and future generations. Providing for human needs by increasing the productive potential through sustainable development and by ensuring equality of economic opportunity for all promoting values that encourage sustainable rates and types of resource utilization and properly managing the waste. Preserving the habitats of other species and ensuring that any use of other species is sustainable; Enabling the people to meet the above conditions in ways of their choosing without hampering the development of others.

Correcting prices, which is the responsibility of markets and governments; Mainstreaming Environmental Considerations; Devising sustainability indicators are to be considered effectively.

SECTION 4

Functional Relation between Environment and Sustainable Development

On the basis of practical approach of the problem it is desirable to develop models on general indicators to access the functional relationship between environment and sustainable development paradigm.

Model 1 : Sustainable development is a function of
SD = (Ep + E - r) (1)

Where increased rate of growth of sustainability (SD), notation of functional relationship between both, environmental needs of the present generation, confined to the inter-generational dimensions (Ep), without compromising the ability of future, generations to meet their environmental needs confined to the intra-generational dimensions, (Ef), reduced rate of population growth®.

Model 2: Ep = (g, H, P, E)

The Present Needs of Environment (Ep) is a function of gross domestic product needs to be accelerated, human resource development index measured by education, health and housing of the society, (H), population control considered to be a major source of environmental hazards, and (P), Energy requirements of the economy considered to be the vital input for development (E).

Model 3: G = (N, P, C)

Gross Domestic Product depends on Employment, Price Stability and Marginal, Propensity to consume is a function of Employment (N), price stability, and (P), marginal propensity to consume ©.

As poverty is the most toxic element of environmental and for its eradication on high rate of GDP is indispensable so that other macro-variables may be useful and conducive for sustainability. Now opportunity of employment that in turn increases the level of MPC of the people at stabilization levels reduces the massive incidence of the most worst environmental pollutant like poverty.

Model 4 : Ef = f (S, W, T, F) The Future Needs of Environment (Ef) is a function of social degradation, purposeful utilization of water resources in various, agricultural and industrial segments (W), transport constraints to environment (V), and utilization of appropriate forest coverage ratio (F).

Model 5 : $-r = f(E_1, H_1, C_1, S_1)$

Reduced rate of Population (-r) is a function of spread of education among the gender sex (E_1), high age for marriage (H_1), development of habit for the use of oral contraceptives (C_1) and standard of living of the people (S_1).

The reduced rate of population is one of the important strategies of sustainable development which cannot be achieved without educating the fair sex, raising the age of marriage or to check the customary of child marriage, free distribution of contraceptives among the urban and rural population and raising standard of living for which they have fascination not to reduce from a certain level.

To sum up for maintaining the equality between environment and sustainable development we must arrive at some feasible conclusions like: When, Ep > Ef = Magnitude of pollution increases which restricts the increased rate of sustainability. When, Ep < Ef = No opportunity of quicker sustainability developed. Therefore, Ep = Ef = A strategy of ecological balance.

Conclusion

The question that is naturally asked is "should we stop all our development activities". The answer is a definite "No". But as stated by the World Commission on Environment and Development, a transition to sustainable development is the only viable path to the new economic era that the world desires. This requires a positive synthesis between economic and environmental policies, programmes and implementations. The transition to the sustainable development mode will require new forms of economic and political relations within the nations and between the nations. Everybody has to realize that this planet of ours is the only known abode of life, and that we have not inherited this from our forefathers, but taken it on loan from the future generations to be returned to them in tact.

Lastly, as a concluding remark, it is nice to mention that the development strategy needs a fresh look. Sustainable development should be our prime goal and for this purpose our developmental strategy must be associated with ecological balances, environmental conditions and pollution free system of economic development, otherwise we will be compelled to have a futureless growth.

REFERENCES

Barry, Dalal-Claton (1992). "Modified EIA and indicators of Sustainability" : First Step Towards Sustainability Analysis; World Bank Conference on Environment and Sustainable Development; Washington, D.C.

Carson, R. (1962). "The Silent Spring", Houghton Mifflin, Boston.

Centre for Science and Environment, Green Economics, Volume 1, No.2, February 1997, pp. 12.

Chatterjee, Biswajit and Karmakar, Asim, (1996). "Environment and Development in an Open Economy", paper published in the seventy-ninth Annual Conference of Indian Economic Association Volume.

Environment (Brundtland Commission) Anil Agarwal (ed.), "State of India's Environment" - A Citizen's Report, 1982 and 1984-85, Centre for Science and Environment, New Delhi.

Furtado, Josel dos R and Tamara Belt with Ramachandra Jammi (2000), "Economic Development and Environmental Sustainability", WBI Learning Series, World Bank, Washington, D.C.

Government of India, Survey of the Environment, 1995, New Delhi, p.32.

Hodges, L. "Environmental Pollution" (2nd ed.), Holt, Reinehardt and Winston, New York, 1977.

Holmbey, J. (1991), "Indicators of Sustainability" Internal memorandum, IIED, London.

IGIDR Series (2000). "Environment, Economics and Development", Environmental Governance-5, Mumbai.

Lester R. Brown and others, "State of the World" (continuing series), World Watch Report (1984 to 1991) W. W. Norton, New York.

Little, IMD and Mirrlus, J.A. (1986, 1990)." Principles for Project Analysis", DAC, OECD, Paris.

Long, R.E. (1989) "The Problem of Waste Disposal", Wilson.

Moran, J.M. (1976), "Introduction to Environmental Science" (2nd Ed.), Freeman, New York.

Perkins, H.C. (1974). "Air Pollution", McGraw-Hill, New York.
Powledge, F. (1982.) "Water, the Nature, Uses and Future of our most precious and Abused Source", Farra Straus, Giroux.
Sastri, M.N. (1993) *History of Planet Earth*, Himalaya Publishing House, Bombay.
Silver, C. S. and R. S. Defries, (1991). "One Earth One Future", National Academy of Sciences, Washington (East-West Press Edition).
Singh, J.K. (1994). "Sustainable Development - A Conceptual Framework", paper presented/delivered at Refresher Course in the Deptt. of Economics, BR.A.B.U.. Muzaffarpur.
Singh, B.N.P. (2001). *Environmental Economics and Development*. pp.120–127.
The Hindu, "Survey of the Environment", 1991 and 1992.
The Hindu. (Daily), October 5, 1994, P.89.
UNCED (1992). *Environmental Management Handbook*, Lewis Publishers, Florida.
World Bank (1991). "Environmental Assessment Source Book" Vols. 1 to 3, The World Bank, Washington, D.C.
World Bank (1992). "World Development Report, 1992", Oxford University Press, New York.
World Commission on Environment and Development (WCED) (1987). "Our Common Future" Oxford University Press, New York.
World Commission on, "Our Common Future", Oxford University Press, New York.

11

MASS MEDIA AND ENVIRONMENTAL SENSITIVITY*

A Case Study of Some of the Reports of Environmental Issues in India

V. Mohanasundaram and *G. Narasimha Raghavan*

"The promotion of environmental awareness and the protection of the environment through investigative reporting has become today a fundamental duty of the media."[1]

Introduction

It would be a truism to say that environmental projects succeed only when there is popular identification with them[2]. Be it a gargantuan 'green technology' project or a puny 'environmental awareness' programme, the role of mass media cannot be obliterated. The basic premise behind such an argument is the 'anticipate and prevent' role of the media[3].

Mass Media

The diffusion of information remains to be the primary task in constructing a 'thinking mass'[4]. Mass media, as the right tool, becomes a crucial communication link between government, citizens, businesses and stakeholders in the environment. Two features predominate the inevitable use of mass media in the propagation of environmental sensitiveness: (a) broad coverage (b) quick reach[5].

---

* Presented at AICTE Sponsored National Level Seminar on "Green Productivity in Small and Medium Enterprises", held at Gandhigram Rural Institute, 5-6 January, 2004.

This brings within the fold of our discussion, the issue of functional/dysfunctional media. With the primary task of 'informing' the people of environmental issues, the mass media is not absolved of its responsibilities.

The derivative effect of this primary task alone will brand it as a functional media, engendering change and integrating public opinion on environmental issues[6]. New media like the Internet and satellite television will have to take up this task along with newspapers, magazines and radio, to disseminate development activities that affect them and the environment[7].

Environmental Sensitivity

The sensitiveness documented here is the positive response of the people to environmental issues. Enhancing such sensitiveness rests with the media. Among the various cliche's, one cliche that would remain faithful to truth is: the environment is an 'in' thing in media[8]. Internationally, it was Rachel Carson's book 'Silent Spring' that elicited a positive response to environment issues by the media[9]. In the domestic arena, one of first and the most *'limelighted'* issue in environment was the threat to the Taj Mahal due to air pollution from a nearby refinery. Any form of environmental sensitiveness evoked by the media would have a two-fold effect (not always mutually exclusive, albeit):

(a) development without affecting environment,
(b) political manipulation and distortions, affecting the environment.

A Case Study of Reportage on Environmental Issues in India

It would not be an ill-considered remark, when it is stated that the magnitude of the environmental issues is a direct resultant of media reportage. Though not all environmental issues can have the same degree of coverage like Chipko, Appiko or Narmda Bachao Andolan (NBA)[10] had, two issues crop up simultaneously:

(a) is it just awareness that the media are spreading? or

(b) are the media providing enough scope for public action against / for environmental issues? With respect to India, can the media in India shed off its 'post-colonial syndrome' of considering environmental issues as neocolonialist intervention?[11] A generic analysis of some of the reports in Indian media on local, regional and national level issues has been undertaken.

- *Local Issue - Bio-medical Waste Management in Coimbatore:* The issue here is the lack of compliance of bio-medical norms despite the passing of the Bio-medical Waste (Management and Handling) Rules, 1998, and the issuance of the order to set-up proper bio-medical waste disposal systems before December 2002. As a local issue, these articles focus on the inertness of the authorities, and the lackadaisical attitude of the hospitals.
- *Regional Issue - Sethusamudaram Project:* This is taken up as an issue of regional importance. Besides, the stake holders include fisher folk and local villagers, not withstanding the effect on the shipping sector.
- *National Issue* - The National Environment Policy 2004 (Internet).

With respect to newspapers, articles in *The Hindu, Dinamani* and *Dinamalar* relating to Bio-medical waste management in Coimbatore have been considered. The first step in analyzing these reports was to identify the promoters of this environmental initiative: Tamil Nadu Pollution Control Board (TNPCB), Coimbatore Municipal Corporation (CMC), Government of Tamil Nadu and the Coimbatore Chapter of Indian Medical Association (IMA). On the other hand, direct beneficiaries have been identified: hospitals and staff, patients and the general public.

The following articles relating to Bio-Medical waste have been chosen:

Table 11.1 : Bio-Medical Waste Management

Sl. No	*Title of t e News*	*Newspaper*	*Date*
1.	4 New Bio-Medical Waste treatment Units Coming Up	The Hindu	19.10.2003
2.	Corporation rejects the disposal of Bio-Medical Waste	Dinamani	06.03.2004
3.	Uncleared Garbage puts hospital in a fix Medical Waste	The Hindu	05.06.2004
4.	Methods laid down for disposal of Bio-Medical Waste	Indian Express	10.06.2004
5.	Bio-Medical waste being disposed of casual	The Hindu	10.06.2004
6.	Epidemic threat looms large over residential colonies	Indian Express	11.06.2004
7.	Private Hospitals doing better in biowaste management	The Hindu	11.06.2004
8.	Contingency plan for bio-medical waste management	The Hindu	13.06.2004
9.	Dumping of biomedical waste persists at compost yard	The Hindu	28.06.2004
10.	Whether the Private hospitals will be watched?	Dinamani	30.06.2004
11.	Burning of Bio-Medical waste in the road will be charged	Dinamani	28.07.2004
12.	Time-bound solution to redress woes of Vellalore residents	The Hindu	04.07.2004
13.	Common facilitator in three months likely	The Hindu	09.07.2004
14.	Training will be given for handling Bio-Medical waste in a proper way	Dinamani	27.07.2004
15.	Place selected in Ukkadam for disposing Bio-Medical waste	Dinamani	28.07.2004
16.	Ukkadam residents oppose choice of site for bio-medical waste disposal	The Hindu	03.08.2004
17.	People opposes for a place selected in Ukkadam for disposing Corporation limit Bio-Medical wastes	Dinamani	03.08.2004
18.	Source Segregation is an essential for Bio-Medical Waste Management	Dinamani	13.08.2004
19.	3,600 tonnes of Bio-Medical waste stock	Dinamani	13.08.2004
20.	Source segregation vital for bio-medical waste disposal	The Hindu	17.08.2004
21.	Common treatment facility in all districts soon	The Hindu	23.08.2004
22.	Reopening incinerator at Chokkampudur considered	The Hindu	24.08.2004
23.	Hospitals to launch time-bound plan for bio-medical waste disposal	The Hindu	26.08.2004

The coverage included :

(i) Meetings - Discussions by the powers that be
(ii) Public Protests
(iii) Efforts of TNPCB, CMC and IMA
(iv) Policies - Setting up of an Incinerator by a private firm.

The major finding was that in spite of good and fair coverage by the print media about the Bio-Medical waste management, which evoked public protests, the authorities (especially TNPCB) were not successful in making all the hospitals to adopt scientific method of Bio-Medical waste disposal.

In the same medium, the Sethusamudaram Project (from *The Hindu and The Business Line),* having a national significance attached to it, was devoted both negative and positive treatment in the national daily. The reports taken into account are :

Table 11.2 : Sethusamudaram Project

Sl. No	*Title of the News*	*Newspaper*	*Date*
1.	Sethusamudaram concerns	The Hindu	20.09.2004
2.	Sethusamudaram Project draws flak from environmentalists	Business Line	30.09.2004
3.	Environmentalists seek wider debate on Sethu Project	The Hindu	22.09.2004
4.	Sethu Project : Avoid ill founded criticism	The Hindu	10.09.2004

Such reports instigated change at three levels :

(a) Understanding of the issue by the masses,
(b) Documenting the view of direct stakeholders like fisher folk and inhabitants,
(c) Survelliance of government action and continuous reporting.

The availability of the National Environmental Policy, 2004 (NEP) on the Internet (www.envfor.nic.in) is an innovative way of reaching the people. This draft NEP was posted in the website to elicit comments and observations. And true to its intention, there were serious review of the draft in many major newspapers and magazines. The ones that are considered here are :

Table 11.3 : National Environment Policy

Sl. No	*Title of the News*	*Newspaper/ Journal*	*Date*
1.	Draft National Environment Policy released	The Hindu	30.08.2004
2.	The National Environment Policy	The Hindu	30.09.2004
3.	National Environment Policy: Ascendance of Economic Factors	Economic and Political Weekly	October 23-29, 2004

This has paved the way for public opinion formation at two echelons :

(a) expert advice on the NEP disseminated through media, and

(b) evaluation of the environmental sensitivity of the National Environment Policy through transparent initiatives.

Conclusion

An insight into such reports has been provided prompted by two objectives: to understand that media can create not only awareness, but also action in the reading population, and to assert that environmental reports have greater impact on the people depending on the scale, region and language of its publication (a local newspaper in a vernacular language can have a greater impact on the people). In so far as environmental issues are concerned, reporting it in any publicity media is not new[12] the idea is to report in such a way that it reeks of authenticity, completeness and forcefulness.

NOTES

1. Appeared in the preface to 'Media and Environment' issue of *Vidura,* 24(1), January - February 1987. The striking feature of this preface is the enduring truth revealed by it, as also its continuing relevance to the present times.
2. The popular identification is qualified by two factors: understanding and

supporting of environmental initiatives. A report 'Environmental Awareness and Education through Mass Media Communication Programme', available at *http://www.odi.org.uk/ tropics/* deals with this phenomenon. The report adds that the most effective way of generating such support is by getting the message across, especially in the case of India's large audience, through the mass media.

3. A hallmark of the media is the freedom given to it to anticipate and write intelligently on potential environmental hazards. Eventually well-informed public opinion arises. However on the front of 'prevention', the media may not be successful in the face of political dysfunctionalism. Despite this, the 'anticipate and prevent' agenda set by the media is a precondition for stopping environmental hazards. *See Bhupal (1987).*
4. The 'thinking mass' used here has the connotation of exposure, assimilation, expression and action, apropos media reports on environmental initiatives.
5. The reasons for such inevitable use of media are (a) accommodation of public opinions in national debates, and (b) allowing public participation in decision making processes in environmental issues. This can be seen in the South Eastern European region as documented in *http://media.rec.orWei background.html.*
6. Additionally, McCormack, remarks, "...(the) function of mass media is to provide both to industry and to the society a coherence, a synthesis, of experience and an awareness of the whole ....". This was quoted in Reddi (1987). This derivative effect is noted further by the author as the surveillance of the environment being one of the fundamental functions of the media.
7. The instigating factors are: (a) sufficient and reliable information, (b) objectivity and (c) media independence.
8. The 'in' thing phenomenon can reduce such environmental initiatives to a ritual and an annual traditional exercise in the case of Annual Environmental Summits. In the bargain, the media reports on environment are even considered 'not news enough'. This is elaborated in a report prepared by Rashmi De Ray for the Environmental Information Systems (ENVIS). A condensed form of the report can be found in Sinha (1986).
9. This was a very influential book published in 1961. This was about the negative effect of DDT use in the USA.
10. Another issue covered under is the 'Silent Valley' report. Many such reports were instigated due to the works of UNDP, UNEP and other international organizations. Largely, it was the NBA issue that prised open the debate on 'development *vs.* environment' in the Indian context. Environmental debate and the cross-flow of information in India are governed by the language in which it is presented and understood, observes Chapman (1997).
11. This is detailed in the book Environmentalism and the Mass Media: The North, South Divide by Graham Chapman, Keval Kumar, Caroline Fraser and Ivor Graber. (London, New York. Routledge, 1997). The skewed attitude of the Indian media is understood from a review of the book by Anantha S. Babbili. This is available at *http://www.wacc.org.uk/modules.php?name=News&file=article&sid=485*

12. It is recorded that Plato had written about deforestation during his time. Even in out epics, there are references to Nature and the environment. The underlying assertion was that environment needs to be safeguarded, and hence these assertions. Alam (2000) gives an ancient and a modern view of environmental issues reportage in the media.

REFERENCES

A. Books /Articles

Anon., (2000), "Mass Media Urged to Devote More to Reporting Environment", available at *http://english.people.com.cn/english/20000706 44794.html*

Anon., "A Need for Assessment of Environmental Journalism in South Eastern Europe", available at *httl2://media.rec.org/ej background.html.*

Alam, Sabiha (2000), "Media and Environment: A Critical Study", in Hussain, Zahid and Vanita Ray (Eds.), *Media and Communication in the Third World,* Kanishka Publishers, New Delhi.

Bhupal (1987), "Environment and the Media", *Vidura,* 24 (1), Jan-Feb, pp.8-9.

Chapman, Graham (1992), "The Mass Media and Global Environmental Learning", available at *http://www.sussex.ac.uk/Units/gec/Tubs/Sublists/Subsasbl.httn*

Chapman, Graham (1997), "Environmentalism, the Mass Media and the Global Silent Majority", available at *http://www.sussex.ac.uk/Units/gec/pubs/briefing/briefs.htm*

Englehardt, Elaine E. and Ralph D. Barney (2001), *Media and Ethics: Principles for Moral Decisions*, Wodsworth, USA.

Gupta, V.S. (1987), "Environmental Awareness and Mass Media", *Vidura*, 24 (1), Jan-Feb, pp.6-7.

Kadambari, Sinha (1986), "Coverage of the Environment in the Print Media", *Vidura*, 23 (3), June, pp.137-139.

Lakshmikantha, M. (2002) "Status of Implementation of Bio-Medical Waste (Management and Handling) Rules, 1998 in the State of Karnataka, *Indian Society of Health Waste Management,* 1 (1), 1" April, pp 38-39.

Mander, Jerry (1996), "The Dark Side of Globalization: What the Media Are Missing", available at *http://www.jzeocities.com/Athens/Acropolis/1232 jzlobal.html*

Mikami, Shunji (1998), "Influence of Mass Media on Public Awareness in Global Environmental Issues in Japan", available at *www.cosmo.soc.toyo.ac.jp*

Nair, S.M. (1987), "The Creation of Environmental Awareness", *Vidura*, 24(1), Jan-Feb, pp.3-5.

Reddi, Usha Vyasula (1987), "Mass Media and Social Structure", *Vidura,* 24 (1), Jan-Feb, pp.31-37.

Rowley, Jack (2002), "Environmental awareness and Socially Responsible Approach—The Ingredients of Continued Success", available at *www.gsmworld.com/using/health.shtml.*

Subramaniam, Meenakshi (1987), "Women and Environment", *Vidura*, 24 (1), Jan-Feb, pp.19-21.

B. Websites

1. 'http://eco.gn.apc.org/
2. http://www.ceeindia.org
3. http://www.odi.org.uk/tropics
4. http://www.recyclingpoint.com.sg/
5. http://www.wacc.org.uk/

12

GREEN PRODUCTION OF JATROPHA BIO-DIESEL OIL THROUGH WOMEN SELF HELP GROUP TO REDUCE RURAL POVERTY AND EMISSION OF DIESEL ENGINE

J. Kavitha, P. Natarajan, S.V. Saravanan and *S. Dhandapani*

Introduction

Green production focuses upon enhancing production without rating adverse impact on the environment. Green production stresses optimum utilization of space manpower, energy raw materials, i.e., all the factors of production. The following flow chart will explain the concept of green production. This paper presents the promotion of green production of Jatropha bio-diesel oil through women self help group to reduce rural poverty and emission of diesel engine. In our country, rural side plenty of waste lands are available. These waste lands can be used for the cultivation of Jatropha plants. The financial expenditure is found to be very much low to cultivate this Jatropha plant. District Rural Development Agency (DRDA) will facilitate the financial and technical activities to promote the Jatropha plant. In recent times due to uncertain supplies and fluctuations in prices for fossil fuel in International Market, the need to search renewable, safe and non-polluting sources of energy assumes top priority. The Jatropha plant is one of the prospective bio-diesel yielding crops. Jatropha Curcas L., is a multipurpose tree of significant economic importance. The following flow chart explain the concept of green production.

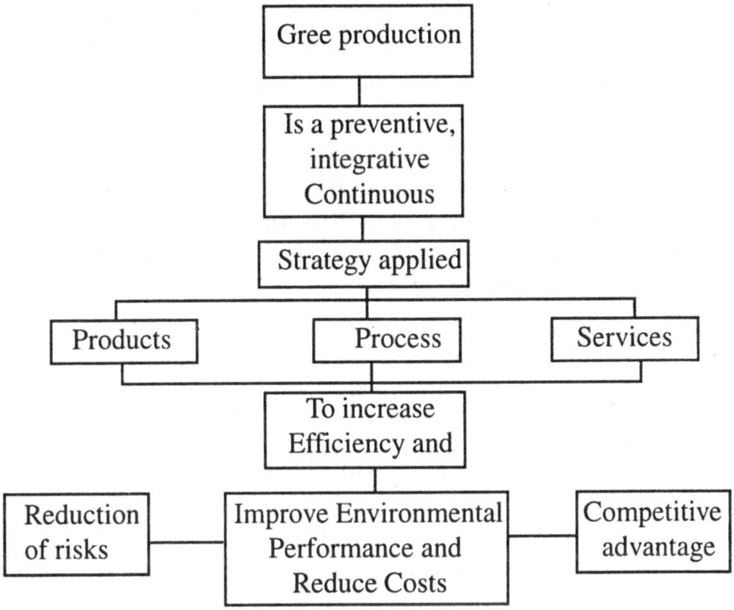

Cultivation of Jatropha through Women Self Help Group (SHG)

Jatropha plant can be cultivated through women self help group. Women can carry out the bulk of the work in agricultural production. Around 70-80 per cent of all fieldwork is done by women whilst most post-harvest and processing tasks are solely their responsibility. All activities involving direct manual labour are assigned to women, Women can play an important role in harvesting and processing of Jatropha plant. About 8-10 per cent women can form a women self help group. They can have their own lands or they can hire the dry lands for lease. For this purpose government is providing loan facilities for SHGs.

Jatropha Curcas

Self-reliance in energy sector is essential for the economic development of a nation. Finding alternative sources of energy, that too renewable, safe and environmentally friendly, assumes top-priority in view of the fast depletion of the fossil fuels. The Bio-diesel produced from the Jatropha oil fills in one such requirement in the recent times.

The oil obtained from Jatropha Curcas seeds is a very good source for the economic production of Bio-diesel. The oil can be processed to give 80 per cent Bio-diesel. The Tamil Nadu Agricultural University at Coimbatore has established a pilot plant for the production of 250 litres of Bio-diesel everyday. Similarly, Mahindra & Mahidra Ltd. have established a 300 tonnes per annum Bio-diesel producing plant at Baroda recently. As such, it is possible to produce very good quality of Bio-diesel from the Jatropha oil obtained from Jatropha Curcas Plant.

Jatropha Curcas also known as physic nut is well adapted to marginal areas with poor soils and low rainfall and can also be grown as a fertile crop. Cattle do not eat the leaves; there is no serious pest or disease attack on the plant. It requires minimum water and be grown as a rain fed crop. It withstands long spells of drought. It requires water only for the first 3 to 4 months after planting for establishment.

This plant can be maintained as a shrub by periodical pruning or can be allowed to grow as tree. It starts yielding after the second year of planting. In a rain fed crop, it can give upto 4000 kgs., seeds per acre from the 4th year by proper crop maintenance. If irrigated once in a week and with proper fertilization and maintenance, the irrigated crop can yield up to 5000 kgs., seeds per acre from the 4th year. This plant goes on yielding upto 35 years after planting, if properly maintained.

Since a tap root is not usually formed when propagated by cuttings, the drought resistance is low; so it is always better to grow this plant from seedlings obtained from seeds of high oil content.

Uses of Jatropha

- It can be grown in waste land and as a fertile crop.

- The oil extracted from the seed is used to produce Bio-Diesel and Glycent Glycerol is used in soap making.
- Glycerol is also used in medicinal, cosmetics and liquor industries.
- The bark of the Jatropha tree is used to manufacture natural dyes for textile industries.
- Jatropha Oil is used for skin disorder and joint paints for both human beings and cattle.
- The oil cake is a very good bio-fertilizer to enrich the soil.
- The leaves are used to grow Tussor Silk worm.

Cultivation Practices

Propagation	:	Through seedlings
Season	:	Can be planted throughout the year. Requires watering for the first 3 to 4 months to get established. Afterwards, if grows with the help of rain water.
Seedlings	:	Two months old seedlings are planted. The stem of the seedlings should be thick, smooth and green. The plant should be healthy.
Spacing	:	For rainfed crop, the spacing between the plants is 5ft. and between the rows is 6 ft., (i.e. 5ft×6ft. spacing). In the case of irrigated crop, the spacing should be 6ft. between plants and 7.5 ft. between the rows. For fencing crop, the spacing should be 2ft. between the plants and 2ft. between the rows. Two rows are planted as fence crop.
Seeding Requirement	:	Per acre Rainfed crop 1225 seedlings Irrigated crop 980 seedlings Fence / Bund crop 418 to 636 seedlings
Planting Method	:	Pits of 30 cm length × 30 cm width ×30 cm depth are dug at the specified spacing. In

that pit add 200gms. of farm yard manure, 300 gms of red earth, plant the seedlings from the polythene bag along with its contents and cover the pit with soil so that the root is fully covered. Immediately irrigate the pit.

Fertilizer : One month after planting apply 20 gms of urea, 120 gms single Super Phospate, 15gms Muriate of Potash per plant and irrigate. In the 3rd month, apply 10 gms. urea per plant and irrigate.

Irrigation : Rainfed crop—No irrigation.
For irrigated crop—Irrigation is required once in a week.

Yield : Per acre approximately.

After Planting	*Rainfed crop*	*Irrigated crop*
First Year	—	Upto 100gms.
Second Year	Upto 1000kgs.	Upto 2000kgs.
Third Year	Upto 1500kgs.	Upto 3000kgs.
Fourth Year	Upto 2000kgs.	Upto 4000kgs.
Fifth Year onwards	Upto 2500kgs.	Upto 5000kgs.

Reinfed-Crop

Income and Expenditure Per Acre

First Year

Sl.No	*Subject*	*Expenditure (in rupees)*
1.	To make pits of 1′×1′×1′ per acre 1225 pits at Rs. 1.50×1225	1,837.50
2.	To fill the pits with manure and sand Rs. 2.00×1225	2,450.00
3.	Price of seedlings Rs. 4.00×1225	4,900.00
4.	Transport and handling Rs. 0.50×1225	612.50
	Total	9,800.00

Second Year to Fifth Year

	Yield in kg	*Income Rs.*	*Expenditure*	*Net profit Rs.*
Second year	1000	5,000	1,500	3,500
Third year	1500	7,500	2,250	5,250
Fourth year	2000	10.000	3,000	7,000
Fifth year	2500	12,500	3,500	9,000

Irrigated Land

Income and Expenditure per Acre
First Year

No.	*Subject*	*Expenditure (in rupees)*
1.	To make pits of 1′ ×1′ ×1′ per acre 980 pits at Rs. 1.50×980	1,470.00
2.	Price of seedlings Rs. 4.00×980	3,920.00
3.	Transport and handling Rs. 0.50×980	490.00
4.	Manure and Sand filling Rs. 2.00×980	1,960.00
5.	Irrigation thrice in a month	1,800.00
6.	Maintenance	1,200.00
	Total	10,840.00

Second Year to Fifth Year

	Yield in kg	*Income Rs.*	*Expenditure*	*Net profit Rs.*
Second year	2000	10000	3500	6500
Third year	3000	15000	4500	10500
Fourth year	4000	20000	5000	15000
Fifth year onwards	5000	25000	6000	19000

This crop can go on yielding for 40 to 50 years.

- At present we buy Jatropha seeds for Rs. 5/- per kg under buy back arrangement, in future we will buy seeds as per the prevailing market price.
- Also we give castor plantation seeds and buy back the same as per the market price.

Applications of Jatropha

The following are the various products derived from the Jatropha plant :

1. Bio-diesel - It is used as a alternative fuel for diesel engine.
2. Bio-oil - It is used to cure skin diseases for humans and animals.
3. Glycerol which is derived from bio-oil is used for soap production.
4. Barks of Jatropha are used in dye production.

Above all these applications, Bio-diesel which is derived from the Jatropha can be directly used as a alternative fuel for the existing diesel engine without making any engine modification. By adopting Transesterification process we can get very good bio-diesel oil which will pollute very less emissions.

Jatropha Bio-diesel Oil Requirements

The following table will give the informanion for the requirement of bio-diesel and the land requirements for the cultivation of Jatropha plant.

Year	*Diesel requirements in million metric ton*	*Bio-diesel @5% in million metric ton*	*Requirement of land in million hectares for the production of 5% Bio-diesel*	*Bio-diesel @ 10% in million metric ton*	*Requirement of land in million hectares for the production of 10% Bio-diesel*	*Bio-diesel @ 20% in million metric ton*	*Requirement of land in million hectares for the production of 20% Bio-diesel*
2004-05	46.97	2.35	1.96	4.70	3.92	9.39	7.84
2005-06	49.56	2.48	2.07	4.96	4.14	9.91	8.28
2006-07	52.33	2.62	2.19	5.23	4.38	10.47	8.76
2011-12	66.90	2.35	2.79	6.69	5.58	13.38	11.19

Source: Report of the committee on development of biofuel, Planning Commission, Govt of India, New Delhi.

Conclusion

There is a plenty of opportunity for women self help group people to cultivate the Jatropha plant in dry lands. The Government and private agencies are providing the necessary technical and financial assistance to SHGs. The cultivation of Jatropha plant through SHGs in large quantity will increase the bio-diesel production. The bio-diesel is used as a alternate fuel for diesel engine to reduce exhaust emissions. The production of bio-diesel will also help to reduce the rural poverty of our nation.

REFERENCES

C.B. Gupta and Dr. N.P. Srinivasan, *Enterpreneurship development in India*, Sultan Chand and sons New Delhi, 1998.

Dom Kundwar, Internal Cobustion Engines, Dhanpat Rai and Sons, Delhi, 1980.

Proceedings of the Enterpreneurship development program, sponsored by Govt. of India, Feb. 24-March 2003.

Proceedings of the One day awareness program on Women entrepreneurship, PSG Polytechnic College, Coimbatore, 2004.

Technical Pamphlet, Renulakshmi Agro Industries (P) Limited, Coimbatore, 2004.

13

ENVIRONMENTAL GOVERNANCE THROUGH RURAL LOCAL GOVERNMENTS

S. Sivakumar and *V.S. Karthik*

Introduction

India faces significant challenges in balancing its increased demand for energy with the need to protect its environment from further damage. Alarming rise of population growth and urbanization make the task all more difficult for the Indian government, as increased vehicular ownership will contribute to existing air pollution problems and urbanization raises the health risks from that pollution. A rapidly increasing population also will lead to an increase in electricity demand, taxing the already straining power generation, transmission and distribution infrastructure and putting pressure on the power sector to add new capacity to the system. Already, shortages of electricity have hampered India's industrial growth. As such electricity is one of the keys to both India's economic development and its environmental protection efforts.

India has made significant efforts in the field of environmental protection developing environmental standards for both products and processes, requiring environmental impact statement in certain areas, and introducing environmental audits. India's strong support of air quality and alternative fuel initiatives has brought progress as well as growing pains to the country. However, in the absence of coordinated government efforts, including strict enforcement, air pollution is likely to continue to worsen in the coming years as urbanization picks up pace and vehicle ownership increases. The UN conference on Human Environment in 1972 proclaimed that the defence and improvement of human environment both natural and man-made had become an imperative good for mankind, to be pursued together with fundamental goals of peace and worldwide economic and social development. The achievement of these goals was the responsibility of citizens, communities, enterprises and institutions at every level. The establishment of viable community based institutions for environment protection was suggested full legal, administrative and technical support. These institutions should be empowered to plan to work out the modality and to implement the various measures for environment protection and conservation.

The seventy-third Constitutional Amendment Act (1992) which came into force from 24th April 1993 lays foundation for strong and vibrant rural local governments, i.e. Panchayati Raj Institutions (PRIs). The rights and duties of the Panchayats are specified in the 11th Schedule of the Constitution which contains 29 lists of subjects. The basic concept of Panchayati Raj is that the villagers should think, decide and act for their own socio-economic interests. The Panchayati Raj Institutions (PRIs) can play decisive role in environment protection and improvement by creating community awareness and providing technical aid on the basis of adaptability, feasibility and sustainability of the necessary conservation measures.

Environmental Governance: The Concept

The concept of "governance" is not new. It is as old as human civilization. Simply put governance means: the process of decision-making and the process by which decisions are implemented (or not implemented).

"Who let this happen? Who's responsible for this mess?" These are typical questions people ask themselves when local environmental disasters happen or when the steady deterioration of the global environment makes the news. For most people, it is not obvious who is "in charge" of the environment, and how decisions are made about developing, using, or managing ecosystems. Governance is about decisions and how we make them. It is about the exercise of authority, about being "in charge." It relates to decision-makers at all levels goverment managers and ministers, business people, property owners, farmers, and consumers. In short, it deals with who is responsible, how they wield their power, and how they are held accountable.

Environmental governance is also about the manner in which decisions are made: In secret or public? Who has "a seat at the table" during deliberations? How are the interests of affected communities and ecosystems represented? How are decision-makers held responsible for the integrity of the process and for the results of their decisions? In fact, governance issues—and matters of environmental governance in particular—are extraordinarily dynamic today. The right of citizen participation; the transparency of organizations and processes; the need to address public corruption; the right to obtain information from governments and businesses about environmental conditions, pollutants, or landuse decisions; the extent to which environmental protection should be included in global trade agreements: All of these are the subject not just of academic policy discussions, but of daily newspaper articles and heated public debate. We see governance at work locally in dicisions about whether we will log or graze a certain area, build a road through a park or a large undeveloped parcel, divert water from a river for farms of houses nearby. These decisions have obvious and immediate environmental impacts.

But governance reaches beyond these high-profile deliberations. It encompasses all the ways in which we exercise authority over the environment more generally, including the timing or overall strategy of management actions such as timber harvests or fishing limits; deciding financing and enforcement; and determining how the benefits from these actions will be allocated. Even the setting of evonomic policies—such as tariffs on imported logs, subsidies for fishing boats of renewable energy, or giving a green light to foreign investment in a natural gas pipeline— is an important aspect of environmental governance, since such policies determine the economic incentives that drive business decisions and influence their environmental and social impacts.

Sometimes we use the term governance very broadly to describe not just the process of decision-making, but the actual management actions—where and when to log, or how to limit fishing or distribute grazing permits—that come from that process. In other words in our day-to-day experience we intertwine environmental governance and ecosystem management, which is where the real impact of decisions becomes visible. In truth, environmental governance goes beyond actual decisions on how to manage natural resources to include the decision-making framework—the laws, policies, regulations, bureaucracies, formal procedures, and codes of conduct— within which managers make their decisions.

Governance and Ecosystems

How we make decisions determines how effectively these decisions are used to manage nature and how fair they are to the people affected. When people participate in nature-related decisions that affect them, they are more likely to support these decisions, and the decisions are more likely to be successully implemented. When they are left out, it is often a recipe for conflict, inequiry, and environmental harm. An ecosystem is a community of interacting organisms and the physical environment they live in. They are the productive engines of the planet—the source of food, water, and other biological goods and services that sustain us. To be effective, environmental governance must lead to fair and sustainable management of ecosystems.

Participation and Environmental Governance

Meaningful participation brings influence. Those who participate in decision-making that affects ecosystems stand the best chance of having their interests represented. Managing ecosystems frequently involves deciding among trade-offs. For example, damming a river may provide farmers more irrigation water but lower the river's fishing potential by interrupting annual spawning runs. Thus, participation by both farmers and fishermen is useful in negotiating how such trade-offs should be made and how the "losers" should be compensated.

Public participation also brings legitimacy, improving the credibility and effectiveness of decision-making processes. Stakeholders can identify conflicts and potential problems that resource managers may have overlooked at an early stage. When all stakeholders have a voice and time is taken to find acceptable solutions, public confidence in the fairness of the decision increases. Especially for large or controversial projects, public involvement in one form or another is required for any broad-based consensus behind the final decision.

Mechanisms to ensure participation in environmental decision-making can take many forms. In democracies, the election or nomination of executives, legislators or other representatives, and the selection of judges provide one way of giving citizens a voice in public policies related to the enviroment. However, the environment is often only one among many concerns of the electrorate, and a candidate's environmental record is seldom the principal factor in determining the outcome of an election. Mechanisms for direct participation are often used to supplement or substitute for elections. For example, in many countries, national policies require that public hearings be held when the environmental impacts of proposed projects are being assessed.

A common challenge to ensuring participation in environmental decision-making is that not all affected stakeholders are equally well-positioned to express their views. For example, at the community level, cultural norms may discourage women from speaking in public, or even from attending a village meeting at which a proposed project is being discussed. Similarly, poorer households may not be able to afford to take time away from productive labour to attend such meetings. More broadly, barriers of distance, language, literacy, and connectivity can also prevent full participation.

Another difficulty is that, even if given the opportunity to participate, stakeholders may lack the capacity to become involved in as meaningful a way as they desire. They may not understand the costs and benefits of the management options or how those options could affect their own interests over time. Or they may not be able to call upon the same sophisticated planning tools or economic analyses that may use to put forward a convincing case.

Decentralization and Environmental Governance

Decentralization—the transfer of powers or responsibilities from a central government to local institutions—goes directly to the question of who gets to make decisions about natural resources. Decentralization can make environmental decision-making more accessible to communities and their representatives, in turn increasing the relevance of those decisions and the likelihood they will be implemented. But decentralization can also occur in ways that leave the *status quo*—central government dominance of decision-making largely unchanged, with little benefit to the environment or local empowerment.

In theory, the poeple who live closest to a natural resource stand to be most affected by its loss or alteration. They have a material interest in managing their environement sustainably. That's why decentralization—the steps that many central government are taking to give regional, municipal, and local institutions responsibility for some public sector functions, from forest management to the provisions of waste disposal services—is an important development in environmental governance.

Recently, several waves of decentralization in developing and developed regions have provided opportunities for local governments to better respond to citizen concerns within frameworks of national environmental and natural resource policy. Central government have often found it hard to enforce some policies—such as grazing allocations, fishing quotas, and restrictions on forest use—because of resistance at the community level to centrally imposed mandates. Under the right conditions, decentralization can bridge this gap by creating ways for local people to negotiate mutually acceptable environmental goals with state authorities.

But decentralization doesn't eliminate the central goverment's role in resource management decisions. Total control over natural resources at the local level is rarely a recipe for environmental success. Communities themselves can decimate resources out of desperation or ignorance, or through corruption or short-term profit-seeking.

Also, while natural resources such as forests and minerals are located in specific communities, their management has wider effects. These include downstream impacts on water supply, regional air pollution, global climate change, and biodiversity loss. Local of communities may overlook these concerns or be incapable of addressing them adequately.

Accordingly, the goal of decentralization should be achieve an appropriate level of local input within a solid framework of national environmental policy: Effective local institutions are needed to negotiate community concerns with national authorities who represent the interests of society at large. Regional and national regulations and democratic processes should also help ensure that all those with a legitimate voice or concern about resouce use get to participate in decision-making. The challenge lies in finding the right mix of local and national powers and responsibilities to achieve sustainability.

The Role of Rural Local Bodies in Environmental Governance

The role of Panchayati Raj Institutions is crucial for environment protection and improvement. Each tier is involved in the process of local planning at their levels. The resolutions are taken by the Gram Sabha are designed and accepted at Gram Panchayat and these plans are examined and prepared by Block Panchayat (or) Panchayati Samiti at intermediate level. At district level, the District Panchayat (or) Zillah Parishad, the District Planning Committee which prepares plans for rational utilization and protection of natural resources. The District Planning Committee forward the plans to the State Government. Such planning process takes care of local conditions, needs of the people, natural resources and geography of the local areas.

The Panchayati Raj Institutions have an effective role in the implementation of various policies, programmes, contributing to environment protection and improvement, designed by State and Central Government. In Tamil Nadu Panchayati Raj Act, it is mentioned that management and maintenance of irrigation tanks in Panchayat area to Panchayat. Besides this, the following roles of Panchayati Raj Institutions are also important:

- Creating will and determination among the rural people to protect and improve the environment :
  * cleaning public roads, public places;
  * polythene-free: village-avoiding polythene bags, plastic cups and plates;
  * discouraging 'pig' rearing;
  * pracicing natural farming method;
  * making the habit to use the pure and safe drinking water.
- Make the people to understand the local situation and dissminate the information about environment :
  * educating the people through rural libraries and public address system;
  * special film shows for protection of environment;
  * informing the people through notice board and using dum-dum;
  * health education;
  * drainage and sewer system.
- Coordinating the activities of various agencies interested in environment protection in rural areas :
  * linkage with non-governmental organizations (NGOs), voluntary organization (VOs);
  * network with youth clubs;
  * creating relationship with other sectoral departments at block and district level.
- Provision of an open forum for discussion on environment protection and improvement :
  * discussing the environmental issues at Gram Sabha meetings;

  * arranging special Gram Sabha and other meetings to discuss environmental issues.
- To give guidance and training to rural people :
  * training the people to cultivate the habit of using toilets instead of open defecation;
  * training the farmers to know about new techniques in agriculture.
- To restrain the proper these activities which are harmful to environment, if necessary by imposing penalty and fine :
  * tree cutting;
  * stealing the sand in irrigation tanks;
  * careless usage of water.

Conclusion

One of the strongest arguments for encouraging better governance is that it requires us to focus on the social dimensions of natural resource use and ecosystem management, in additon to the technical details of how to manage. This includes how we value ecosystems, how we set management goals, how we negotiate trade-offs between our decisions—including impacts on the poor—are equitably shared. In fact, a focus on governance adds an explicit consideration of fairness to the goals of ecosystem management.

Science and technology can help us answer questions about what kinds of management actions are most effective in protecting or restoring ecological integrity. For example, conservation science can estimate how large an area of forest we should preserve to ensure the survival species of wildlife or plants. Atmospheric science can model how quickly greenhouse gas emissions must be reduced to stabilize their build-up in the atmosphere and avoid catastrophic changes in the global climate system.

However, conservation science cannot tell us how best to resolve conflicts between local communities and logging companies over the fate of a forest, and atmospheric science cannot tell us how responsibility for reducing emissions should be distributed. These are governance questions, involving the balance of ethical and moral concerns, social and economic goals, and the capacities of natural systems.

Similarly, economic analysis can answer questions about the most efficient methods for achieving various ecosystem management objectives. For example, economic analysis can inform the design of a system of taxes and subsidies to encourage electricity producers to build more efficient power plants, or to encourage polluting factories to reduce their emissions.

But economic analysis cannot tell us how best to respond to community concerns over the siting of those power plants and factories. Again, it is the challenge of governance to answer questions such as "What's fair?" "What's the right balance?" and sometimes "Who benefits and bears the consequences?" —in addition to giving insights into what is efficient and effective in the real world of competing interests.

REFERENCES

D.D. Basu, *Constitution Law of India*, New Delhi.

George Mathew (ed), (1995), *Status of Panchayati Raj in the States of India 1994*, Concept Publishing Company, New Delhi.

N.R. Inamdar (1993), Sustainable Development: Concept and Policy, *Indian Journal of Public Administration*, Delhi, Vol. XXIX, No.3.

Sukhla, S.K. (1989), *Environmental Pollution and the Law*, Sashi Publications, Jaipur.

The Constitution (Seventy Third Amendment) Act, 1992 published in the Gazette of India dated 20th April, 1993.

14

TOURISM AND ENVIRONMENTAL ISSUES

M. Soundarapandian and *T. Saravanan*

Introduction

Tourism has emerged as the largest industry of the 20th century and is projected to grow even faster in the 21st century. International tourism is the world's largest export earner and vital to the balance of payments of many countries. For small developing nations tourism is often the only way they can compete in the dynamically expanding service sector.

The number of world arrivals was increased by 38.3 per cent when it compared with the year 1992. The India's arrivals were also shows the nearest increase while comparing the world arrivals.

The environment and local culture receive a boost when authorities restore monuments, open museums and establish natural parks to lure visitors. As tourism increases, so does a destination's need to improve infrastructure to handle the influx. The quality of the environment, both natural and man-made, is essential to tourism. However, tourism's relationship with the environment is complex. It involves many activities that can have adverse environmental effects. Many of these impacts are linked with the construction of general infrastructure such as roads and airports, and of tourism facilities, including resorts, hotels, restaurants, shops, golf courses and marinas. The negative impacts of tourism development can gradually destroy the environmental resources on which it depends.

On the other hand, tourism has the potential to create beneficial effects on the environment by contributing to environmental protection and conservation. It is a way to raise awareness of environmental values and it can serve as a tool to finance protection of natural areas and increase their economic importance.

Negative impact from tourism occur when the level of visitor use is greater than the environment's ability to cope with this use within the acceptable limits of change. Uncontrolled conventional tourism poses potential threats to many natural areas around the world. It can put enormous pressure on an area and lead to impacts such as soil erosion, increased pollution, discharges into the sea, natural habitat loss, increased pressure on endangered species and heightened vulnerability to forest fires. It often puts a strain on water resources, and it can force local populations to compete for the use of critical resources. In this paper an attempt has been made to analyse the significance of tourism and its impact on the environment.

Tourism Impact on Natural Resources

Tourism development can put pressure on natural resources when it increases consumption in areas where resources are already scarce.

Water Resources

Water, and especially fresh water, is one of the most critical natural resources. The tourism industry generally overuses water resources for hotels, swimming pools, golf courses and personal use of water by tourists. This can result in water shortages and degradation of water supplies, as well as generating a greater volume of wastewater.

Local Resources

Tourism can create great pressure on local resources like energy, food, and other raw materials that may already be in short supply. Greater extraction and transport of these resources worsen the physical impacts associated with their exploitation. Because of the seasonal character of the industry, many destinations have ten times more inhabitants in the high season as in the low season. A high demand is placed upon these resources to meet the high expectations tourists often have proper heating, hot water, etc.

Land Degradation

Important land resources include minerals, fossil fuels, fertile soil, forests, wetland and wildlife. Increased construction of tourism and recreational facilities has increased the pressure on these resources and on scenic landscapes. Direct impact on natural resources, both renewable and non-renewable, in the provision of tourist facilities can be caused by the use of land for accommodation and other infrastructure provision, and the use of building materials. Forests often suffer negative impacts of tourism in the form of deforestation caused by fuel wood collection and land clearing.

Tourism and Pollution

Tourism can cause the same forms of pollution as any other industry air emissions, noise, solid waste and littering, releases of sewage, oil and chemicals, even architectural/visual pollution.

Air Pollution and Noise

Transport by air, road, and rail is continuously increasing in response to the rising number of tourists and their greater mobility. Transport emissions and emissions from energy production and use are linked to acid rain, global warning and photochemical pollution. Air pollution from tourist transportation has impacts on the global level, especially from carbon dioxide (CO_2) emissions related to transportation energy use. And it can contribute to severe local air pollution. Some of these impacts are quite specific to tourist activities.

Noise pollution from airplanes, cars, and buses, as well as recreational vehicles such as snowmobiles and jet skis, is an ever-growing problem of modern life. In addition to causing annoyance, stress, and even hearing loss for it humans, it causes distress to wildlife, especially in sensitive areas. For instance, noise generated by snowmobiles can cause animals to alter their natural activity patterns.

In areas with high concentrations of tourist activities and appealing natural attractions, waste disposal is a serious problem and improper disposal can be a major despoiler of the natural environment—rivers, scenic areas, and roadsides. In mountain areas, trekking tourists generate a great deal of waste. Tourists on expedition leave behind their garbage, oxygen cylinders and even camping equipment. Such practices degrade the environment with all the detritus typical of the developed world, in remote areas that have few garbage collection or disposal facilities.

Sewage

Construction of hotels, recreation and other facilities leads to increased sewage pollution. Wastewater has polluted seas and lakes surrounding tourist attractions, damaging the flora and fauna. Sewage pollution can threaten the health of humans and animals.

Aesthetic Pollution

Often tourism fails to integrate its structures with the natural features and indigenous architectural of the destination. Large, dominating resorts of desparate design can look out of place in any natural environment and may clash with the indigenous structural design.

A lack of land-use planning and building regulations in many destinations has facilitated sprawling developments along coastlines, valleys and scenic routes. The sprawl includes tourism facilities themselves and supporting infrastructure such as roads, employes housing, parking, service areas, and waste disposal.

PHYSICAL IMPACTS

Attractive landscape sites, such as sandy beaches, lakes, riversides, and mountain tops and slopes, are often transitional zones, characterized by species-rich ecosystems. Typical physical impacts included the degradation of such ecosystems. Physical impacts are caused not only by tourism-related land clearing and construction, but by continuing tourist activities and long-term changes in local economies and ecologies.

Physical Impacts of Tourism Development

Construction Activities and Infrastructure Development

The development of tourism facilities such as accommodation, water supplies, restaurants and recreation facilities can involve sand mining, beach and sand dune erosion, soil erosion and extensive paving. In addition, road and airport-construction can lead to land degradation and loss of wildlife habitats and deterioration of scenery.

Deforestation and Intensified or Unsustainable Use of Land

Construction of ski resort accommodation and facilities frequently requires clearing, forest land. Coastal wetlands are often drained and filled due to lack of more suitable sites for construction of tourism facilities and infrastructure. These activities can cause severe disturbance and erosion of the local ecosystem, even destuction in the long-term.

Marina Development

Development of marinas and breakwaters can cause changes in currents and coastlines. Furthermore, extractions of building materials such as sand affects coral reefs, mangroves, and hinterland forests, leading to erosion and destruction of habitats. In the Philippines and the Maldives, dynamiting and mining of coral for resort building materials has damaged fragile coral reefs and depleted the fisheries that sustain local people and attract tourists.

Trampling

Tourists using the same trail over and over again trample the vegetation and soil, eventually causing damage that can lead to loss of biodiversity and other impacts. Such damage can be even more extensive when visitors frequently stray off established trails.

Tourism Contribution to Environmental Conservation

The tourism industry can contribute to conservation through :

Financial Contributions

Tourism can contribute directly to the conservation of sensitive areas and habitat. Revenue from park-entrance fees and similar sources can be allocated specifically to pay for the protection and management of environmentally sensitive areas. Special fees for park operations or conservation activities can be collected from tourists or tour operators.

Contributions to Government Revenues

Some governments collect money in more far-reaching and indirect ways that are not linked to specific parks or conservation areas. User fees, income taxes, taxes on sales or rental of recreation equipment, and licence fees for activities such as hunting and fishing can provide governments with the funds needed to manage natural resources. Such funds can be used for overall conservation programnes and activities, such as park ranger salaries and park maintenance.

Improved Environmental Managerial and Planning

Sound environmental management of tourism facilities and espacially hotels can increase the benefits to natural areas. But this requires careful planning for controlled development, based on analysis of the environmental resources of die area. Planning helps to make choices between conflicting uses, or to find ways to make them compatible. By planning early for tourism development, damaging and expensive mistakes can be prevented, avoiding the gradual deterioration of environmental assets significant to tourism.

Environmental Awareness Raising

Tourism has the potential to increase public appreciation of die environment and to spread awareness of environmental problems when it brings people into closer contact with nature and the environment Tills confrontation may heighten awareness of the value of nature and lead to environmentally conscious behaviour and activities to preserve the environment.

The tourism industry can play a key role in providing environmental information and raising awareness among tourists of the environmental consequences of their actions. Tourists and tourism-related businesses consume an enormous quantity of goods and services; moving them towards using those that are produced and provided in an environmentally sustainable way, from cradle to grave, could have an enormous positive impact on the planet's environment.

Protection and Preservation

Tourism can significantly contribute to environmental protection, conservation and restoration of biological diversity and sustainable use of natural resources. Because of their attractiveness, pristine sites and natural areas are identified as valuable and the need to keep the attraction alive can lead to creation of national parks and wildlife parks.

Alternative Development in Tourism

Tourism can provide an alternative to development scenarios that may have greater environmental impacts, Eco-Tourism.

Eco-Tourism

Eco-tourisin has attracted increasing attention in recent years, not only as an alternative to mass tourism, but also as a means of economic development and environmental conservation. Being a nature based tourism; it takes into account the natural ecological attraction, their conservation and development. Its main aim is to safeguard the environment, making it beneficial to the local people by generating revenue and education and pleasure for the tourist.

The market for nature holdays is certainly a growing one. In 1993, the World Tourism Organization (WTO) has estimated the nature tourism generates 7 per cent of all international travel expenditure, where the total annual global earning from tourism is nearly $3 trillion and 212 million people are employed (Eagle, 1997). Year 2002 has been declared as a International Year of Eco-tourism (IYE) by the United Nations. This reflects its global importance.

The following points should be considered while implementing eco-tourism :

- The scale of development should be appropriate for the local community and for the capacity of the environment to support the facility.
- Zoning measures should be considered to reduce the threat to the environment.
- Roads or any trails should be placed in a manner so as to minimize intrustion on the environment.
- Measures should be taken during construction to minimise negative effects on the environment.
- Construction of the facility should use locally available material and blend with the surroundings.
- Energy sources should be environmentally sound and suitalbe.
- Appropriate technologies should be employed for the treatment of organic and other wastes.
- Local communities should be involved in planning, construction and operations.

Conclusion

Regulatory measures help offset negative impacts, for instance, controls on the number of tourist activities and movement of visitors within protected areas can limit impacts on the ecosystem and help to maintain the integrity and vitality of the site. Such limits can also reduce the negative impacts on resources. Limits should be established after an in-depth analysis of the maximum sustainable visitor capacity.

REFERENCES

Andrew Holden and Mark Ewen, 'Issues of Natural Resource Usage, Social Well-Being and Tourism in the Annapurna Area of Nepal' – *Asean Journal on Hospitality & Tourism*- Jan-July, Vol-3, Issue 1, 2003.

Bet Lagarense, "Community based Eco-tourism Development to Reduce Climate Change and Man Made Coral Bleaching : The Case of Bunaken National Park"— *Asean Journal on Hospitality & Tourism*, Index Volume 2, 2003.

Jinyang Deng, "Eco-tourism, Protected Areas, and Globalization : Issues and Prospects in China"- *Asean Journal on Hospitality & Tourism*—Index Volume 2, 2003.

Nick Kontogeorgopoulos, "Towards a Southeast Asian Model of Resort-Based" "Mass Ecotourism : Evidence from Phuket, Thailand and Bali, Indonesia", —*Asean Journal on Hospitality & Tourism* –Index Volume 2, 2003.

Wu, C.-T. "Issues of tourism and Socio-economic Development." *Annals of Tourism Research* 9(3): 317-330, 1982.

15

HEALTH HAZARDS DUE TO CONVENTIONAL ENERGY AND ENVIRONMENTALLY SUITABLE RURAL ENERGY SYSTEMS

M. Soundarapandian and *A. Inigo Antony Beaula*

In India, expecially in rural areas, more than 80 per cent of the population depends upon conventional (or) traditional energy sources for meeting their cooking energy requirements. Because of traditional (low efficient) *chula*, partial ignition of the biomass fuels takes place. So, large quantity of smoke is emitted. Hence indoor air pollution is caused.

Cooking activities are done only by women. So, they are exposed to indoor air pollution and they have to inhale the smoke which contains hazardous pollutants. Hence premature death of women and children is caused. Exposures of indoor air pollution leads to other diseases like difficulty in breathing, headache, eye-irritation and lung cancer, asthma, turberculosis, cataract, hypoxia and low birth weight.

The main reasons for these health hazards are analysed on the basis primary data collection. The main objective of this paper is :

- To find out the energy sources used for meeting the cooking requirements.
- To identify the health hazards caused by using conventional or traditional energy sources.
- To suggest the suitable alternatives for meeting the rural energy requirements.

In Dindigul block, 3 revenue villages, namely Mullipadi, Seelapadi and Balakrishnapuram have been selected and "study area". Identification of "study area" is based on the agricultural developments and developments in production activities, improvements in adopting remarkable energy sources, positive economic impacts and environment, economic status of households and population pressure.

A total of 75 respondents (25 resopndents from each village) have been selected and surveyed. Respondents are classified into 5 categories, namely *Large Farmer (LF), Small Farmer (SF), Marginal Farmer (MF), Agricultural Landless Labourers (ALL) and others.* Farmers who are having 5 and above 5 acres are considered large farmers. Farmers having 2.5 to 5 acres are considered small farmers. Marginal farmers are having 2.5 and below 2.5 acres. Agricultural landless labourers indicate the workers without land, working as "*Coolie*" in other agricultural land works. "*Others*" indicates the workers engaged in trade and commerce, industrial units, rural artisans, services and other community services.

Sources

Conventional Energy used by the Households are Firewood, Kerosene, LPG, Biogas, Electricity. The sources of the energy used by the sample households are analysed in Table 15.1.

Table 15.1 : Classification of Conventional Energy Consumers

Sl.No.	*Energy Sources*	*No. of Respondents*	*Total*
1.	Fire wood	63(84.00)	75(100)
2.	Kerosene	70 (93.33)	75(100)
3.	LPG	25(33.33)	75(100)
4.	Biogas	5(06.67)	75(100)
5.	Electricity	65(86.67)	75(100)
	Total	**228(60.80)**	**375(100)**

The term *Biomass* includes all plants like trees, agricultural plants, bush, grass and algae, and their residues after processing. The direct way of using a biomass resource is to burn it, the fuel most commonly used being wood. In rural areas of India, a substantial part of cooking energy requirements of millions of village households are met by burning wood. In these three selected villages, *84 pre cent (63 respondents) of the people depend on firewood* for their cooking energy requirements.

Kerosene is one of the main sources of energy, which satisfies the cooking and lighting energy requirements of the rural people. In the "study area" *93.33 per cent (70 respondents) are using kerosene* for cooking and lighting.

L.P.G. is an another source of energy used for cooking purposes. Out of 75.3 respondents, *25 respondents are using L.P.G.* In these study areas, most of the people are using only firewood. Hence for meeting their cooking requirements, most of their time will be spent of fuelwood collection. It these people are provided with the superior energy sources like L.P.G., they will get more time to spend with their family, and for education and craft. So, they will get an additional suource of earning. Ultimately, their family status will be improved.

Biogas is obtained through the anaerobic fermentation of wet livestock wastes. It is a mixture of CH_4 (45 to 70 per cent) and CO_2. Biogas is a clean, smoke and soot-free fuel. In these selected revenue villages, out of 75 respondents *only 5 respondents are using Biogas.* Two types of biogas plants are used in the study area. *One floating drum type and other fixed dome type.* Because of the problems and recurring maintenance cost of the former, the latter is widely used in rural areas.

Energy use in domestic lighting includes consumption of energy for lighting, radio, television and fans wherever possible. Of course, purpose of energy use is limited and in small quantity in the cases of radio, television and fans. *Main sources of domestic lighting are electricity, kerosene in the region.* Among them *predominant sources of lighting is electricity followed by kerosene.*

Among 75 respondents, *65 respondents have been using the electricity for lighting.* In these selected revenue villages, 3 alternative sources of fuels for lighting have been seen, viz., (i) using electricity only, (ii) using kerosene only, (iii) using electricity, and kerosene both. These alternatives vary across socio-economic or occupational status of the respondents.

Health Hazards

Due to the use of conventional (or) traditional energy sources, the respondents from this "study area" are affected by major health hazards like *Cancer, Eye-irritation, Headache, Respiratory problems such as Tuberclosis, and Cough.* The findings of the study is given in Table 15.2.

Table 15.2 : Health Maladies Analysis

Hazards	*LF*	*SF*	*MF*	*All*	*Others*	*Total*
Cancer	–	–	–	1	–	1
	(0.00)	(0.00)	(0.00)	(6.67)	(0.00)	(1.33)
Respiratory	3	5	4	3	5	20
problems	(2.00)	(33.3)	(26.7)	(20.0)	(33.3)	(26.7)
Eye-irritation	8	7	6	5	6	32
(Cataract)	(53.3)	(46.7)	(40.0)	(33.3)	(40.0)	(42.7)
Headache	4	3	5	6	4	22
	(26.7)	(20.0)	(33.3)	(40.0)	(26.7)	(29.3)
Total	15	15	15	15	15	75
	(100.0)	(100.0)	(100.0)	(100.0)	(100.0)	(100.0)

Source : Computed from primary data. Figures within parenthesis denotes percentages.

Despite many reasons for the health problems in rural areas, use of traditional energy also creates the health hazards. Among 75 respondents, only one respondent is affected by Cancer, and 20 respondents (26.7%) are having respiratory problems like asthma and cough. 32 respondents (42.7%) are affected by eye-irritation problems such as cataract and 22 respondents (29.3%) are having headache. Partial ignition of biofuels leads to the emission of hazardous pollutants like *Carbon monoxide, Carbon dioxide, Nitrogen oxide and Sulphur dioxide*. The stress of these toxic gases is manifested in the form of headache, dizziness, respiratory problems, eye-irritation and cancer. All these problems lead to Psychosomatic illness.

Reasons for Health Hazards

Reasons for the Health Hazards are analysed from the respondent's response. Five major reasons are found out from the responses. They are *Smoke, Dust, Traditional chulas, Illiteracy, Low income profile of the respondents and unawareness of the respondents* regarding highly efficient renewable energy sources.

Table 15.3 : Causes For Health Problems With Percentage Points

Reasons	*LF*	*SF*	*MF*	*ALL*	*Others*	*Total*
Smoke, dust	2 (13.33)	3 (20.00)	3 (20.00)	3 (20.00)	4 (26.67)	15 (20.00)
Traditional *Chula*	2 (13.33)	3 (20.00)	2 (13.33)	2 (13.33)	2 (13.33)	11 (14.67)
Illiteracy	5 (33.33)	2 (13.33)	3 (20.00)	3 (20.00)	1 (6.67)	14 (18.67)
Low income	2 (13.33)	2 (13.33)	4 (26.67)	4 (26.67)	4 (26.67)	16 (21.33)
Unawareness	4 (26.67)	5 (33.33)	3 (20.00)	3 (20.00)	4 (26.67)	19 (25.33)
Total	15 (100.0)	15 (100.0	15 (100.0)	15 (100.0)	15 (100.0)	75 (100.0)

Source : Computed from primary data. Figures within parenthesis denotes percentages.

Among 75 respondents, 15 respondents (20%) say that 'Smoke and dust' is one of, the reasons for the health hazards. Due to the partial ignition of the firewood, large quantity of smoke is emitted. 11 respondents (14.67%) realise that traditional *chula* is the other reason for the health problems. When the traditional *Chula* is used for cooking purposes, there is no separate chimney for eliminating the smoke. So, the consumer has to inhale the smoke emitted from the *Chulas*. Inhaling this smoke is equivalent to smoking 20 packets of Cigarettes per day. So, the traditional *chula* is the other reason for the health problems.

Among the 75 respondents, 14 respondents (18.67%) say that illiteracy is the other reason for the health hazards. Because of illiteracy, many people don't know anything about the renewable energy sources and their applications. So they blindly depend upon natural resources such as firewood, and agro-residues for meeting the cooking energy requirements. So, it indirectly links with deforestation, and desertification. Conventional fuel users are also affected by many other health problems. So, the illiteracy is one of the reasons for the health hazards.

Another important reason is low income profile of the people. This problem is pointed out by 16 respondents (21.33%). The cost of superior energy sources like LPG and steam cooking (Steamer) is high. So many rural population living below the poverty line couldn't adopt superior or renewable energy sources for meeting their energy requirements. So, these people have to depend upon the traditional (or) conventional energy sources even though it causes many diseases. The last but not the least reason is unawareness of the people.

Among the 75 respondents, nearly 14 respondents are having sophisticated life. Other respondents are middle income group and 20 respondents are living below the poverty line. Nearly 25 respondents are having food twice a day. So, they are not aware of the renewable energy sources and devices. Even though many banks come forward to give soft loans to the people, they do not utilise those opportunities. People don't know about the subsidy on the installation of renewable energy devices.

Unawareness regarding these renewable energy sources increases the usage of traditional on conventional energy sources and it indirectly increases the health hazards also. When the rural population adopts these renewable energy sources, it reduces the time taken for cooking food. So, the people will have lot of time to spend for education, craft and in finding all additional sources of income. So, these private benefits will reach the people only, when energy awareness programmes are conducted by the government. So, this unawareness of the people is another reason for the health hazards.

Suggestion

The main suggestion is to foster socially and environmentally appropriate rural energy systems. The following suggestions are some of the possibilities enhancing the development of *Fuelwood Plantations, Biogas, improved for efficient devices (Improved chulas) and Solar cookers.*

Fuelwood Plantations

In this Scheme, selected species of trees are planted and harvested over regular intervals of time in a phased manner. So that wood is continuously available for cooking or allied purposes. In order to augment the production of *fuelwood, the Fuelwood Study Committee (Planning Commission, 1982)* suggested three feasible approaches. These were (1) Plantation of farmlands, (2) Plantation on barren and wasteland, and (3) Plantation of degraded forest.

Plantation of farmlands can be done by planting fast growing, non-shady plants around the farm boundary in such a way that the required fuelwood could be harvested annually. In this approach, 50 trees could be planted in one acre of land and 10 trees could be harvested each year.

Plantation on barren and wasteland is defined as land which is either not available for cultivation or is left out of cultivation due to some reason or the other. But, this second approach results a reduction of employment opportunities, by converting the agricultural land into wood plantation, and this will result a negative impact on rural poor. If the government comes forward to buy wastelands from the big land holders and lease them to the landless or marginal farmers with an agreement of energy plantation, these problems can be easily solved.

Biogas

- Biogas is a clear burning smoke and soot free fuel. It is cheap and extremely convenient, and doesn't cause air pollution. It reduces the kitchen related health problems.

Use of biogas also reduces the drudgery of women in gathering fuelwood. Thus, biogas users an all around improvement in the quality of life in the household.

- The manure produced through a biogas unit has a comparative advantage over ordinary manure in terms of manorial values. The nitrogen content of cattle dung is conserved in a biogas unit and to the advantage, part of the nitrogen is converted to ammonical form. It is good source of micro nutrients like *Zn, Fe, Mn, Cu* which are fast decreasing in many soils. (Dr. R. Kasturibai, 2003).
- (Based on the effective heat produced, 2 cum biogas plant could replace in a month, fuel equivalent of 26 kg of LPG or 37 litres of kerosene or 88 kg of charcoal or 210 kg of fuelwood or 740 kg of animal dung. 1 cum biogas can generate 1.25 kw electricity, can run a 1 hp motor for 2 hours, cart drive a 3-tonne lorry to 2.8 km. (Dr. R. Kasturibai, 2003).
- The other advantage is that, biogas plants are decentralized systems and they don't require huge capital investments and elaborate distribution networks. They can be installed even in remote areas with very low investments.

However, Biogas plant has some vexing problems such as lower gas yield in colder claimates, the non-availability of easy to operate, lack of space. But, development of biogas plant is an important route to the solution of fuel shortage problems.

Improved *Chulas*

Firewood, agro-residues, cattle-dung cakes are the main source of domestic energy for cooking and heating activities in rural areas. If they use improved *chulas*, the following benefits will be attained.

- Satisfactory combustion of fuel.
- Reasonable cost of the stove including chimney and control of the draft by a damper.

- Lesser amount of smoke is produced, and it is eliminated by the suitable chimney.
- Cleaning the *chulas* is very simple and conventional
- Time spent on cooking is very less.
- Ignition of the improved *chulas*, doesn't produce any irritations.
- Lesser amount of firewood consumption.

Acceptability of improved *chulas* is determined by many factors such as *cooking efficiency, cost, size and type of fuel available, family size, cooking practices and types of food to be cooked and space available* for cooking activities. All these factors vary from region to region. Improved *chulas* helps to conserve fuelwood. So, it indirectly reduces deforestation. So, the government should take proper steps to popularise it.

Solar Cookers

Solar cooker is another alternative for cooking energy requirement Temperatures around 100°C can be obtained in these cookers on sunny days and pulses, rice, vegetables etc. can be readily cooked. The time taken for cooking depends upon the solar radiation and varies from ½ to 2½ hours. In most places in India, solar cookers can be used for at least 250 days in the year. Because of the relatively low payback period, solar cookers have become fairly popular in many states.

Conclusion

This study gives a detailed report of environmental and health impacts against using conventional (or) traditional biomass fuels and striking with force on other alternatives available for meeting the energy requirements. The awarness regarding health impacts against conventional sources and collision with force on the other alternative renewable energy sources will reach the people's mind, only if the Government takes proper steps to spread this awarness by conducting many development programmes about renewable energy sources. They must try to increase the subsidy on the installation cost of renewable energy devices. Only then will atleast the middle income, high income group come forward to adopt these renewable energy sources which lead to sustainable economic development of our country. Adopting a proper renewable energy source is a important way of reducing health related problems and keeping the environment clean and safe.

REFERENCES

Agarwal, A.A., Forgotten energy crisis, *New Scientist* 97 (1344), 1983 pp. 377 - 379.

Bowonder, B., Deforestation in India, *International Journal of Environmental Studies*, 18 : 1986, pp. 223-236,

C. Moorthy, "Indian Scenario" *Yojana*, November 16-30, 1990. pp. 4-6.

Cooking with biofuels : Health impact on rural women, *Economic and Political Weekly*, Vol.38, No. 26, 2003, June 28. pp. 2681-2692.

Dhulasi Birundlia Varadharajan, *Energy Economics*, Sterling Publishers, New Delhi, 1992, pp. 76-78.

D.B. Kannade, P.S. Yadar, H.G. More and S.M. Pokharkar, "Briquetting : A sure way to solve problem of Fuel Wood in Rural Areas, Khadigramodyog, the *Journal of Rural Economy,* Vol. 40, No. 8, 1991, July, pp. 399 to 400.

Hema Patel, "*Energy* for *Cooking*", Gujarat Home Energy Conservation Development Agency, Vol. 2,1989, December, pp. 56-57.

Kasturibai R. *Biogas Technology for Rural Development,* paper presented at a state level workshop on Renewable Energy for Sustainable Development, 2003, March 15, 16.

Muthuchelian K., *Energy Plantation for Power Generation and Rural Energy Supply*, paper presented at a state level workshop on Renewable Energy for Sustainable Development, 2003, March 15,16,

Parikh, K., *Household Energy Assessment, Integration of Approaches and Additional Factors*, Biomass Elsevier Applied Science Publishers, Ltd., England, pp. 73-84.

Solomon, Raj D., *Impact of improved smokeless chulas on Energy use pattern of Rural households of Udankudi Block* in Tuticorin District, paper presented at a state level workshop on renewable energy for sustainable development 2003, March 15, 16.

16

ROLE OF MANAGEMENT IN ENVIRONMENTAL ISSUES

Mr. V. Sudhakar and *M.S.K.N. Nagendran*

Introduction

The world of environment is the combination of atmosphere, hydrosphere, lithosphere and bio-sphere. Managing environmental and safety issues is challenging, time-consuming and expensive. Being responsible for these matters is more than just saving money and protecting the company. The natural environment is clean but due to the multifarious activities of man it gets polluted resulting in what is called as environmental pollution. In normal circumstances it remains clean it is enjoyable otherwise the human being will become an endangered social animal because of the pollution.

In recent times, environment has emerged as a major world-wide concern. Pollution in particular is perceived as a serious threat in industrialized nations, while natural resource degradation due to pollution and population explosion is becoming a serious impediment to the economic development in developing countries.

Thc pollution of various resources has gone to such an extent that we are unable to breathe fresh air and drink fresh water. On one hand the advancements of science and technology have added to the human comforts by giving us automobiles, electrical appliances, supersonic jets, space crafts, better medicines, better chemicals to control harmful insects and other pests, etc., but on the other hand, they have given us a very serious problem to face that is pollution.

Pollution can be defined as an undesirable change in the physical, chemical or biological characteristics of the air, water or land that can harmfully affect health, survival or activities of humans or other living organisms. Pollution does not have to cause physical harm, but it may merely interfere with human activities. Chemical spewed into the air or water from an industrial plant may be harmful to humans and other organisms living near the plant. If expensive pollution controls are required, the plant may be forced to shut down.

Problem in India

India faces serious problems of environmental pollution from untreated community and industrial wastes flowing into our rivers and streams, increasing use of fertilizers and pesticides, dumping of organic wastes etc., are issues that need to be tackled on the priority. Air pollution, long regarded as the bane of industrialized nations looms large over our country.

The developing countries, which have 80 per cent of the world population, are bothered and burdened with adverse affects on environment. During the sixties people all over the world became more concerned about the quality of life. The miracles of sixties have become the nightmares of the nineties.

Throughout the history of man-kind, it is evident that there has been some awareness of the environment and the need for its conservation. But unfortunately the awareness began to decline gradually with the advancement of industrialization and urbanization with over conceptive lifestyles. Hence, the need of management in environmental issues has taken place at this juncture.

The most common industry that everyone of us knows and quite related with us is transportation. Take for example this transportation industry and its effects on the green globe.

Motor vehicles emit 4,000 million tonnes of carbon dioxide each year. That's one quarter of the world's total carbon dioxide output. This carbon dioxide is the main cause of global warming.

If global warming continues the polar ice caps will melt and many places will be under water, e.g. London.

Each litre of petrol burnt produces about 2.4 kg of carbon dioxide regardless of size, make, shape, colour, fluffy dice or go faster stripes.

Each litre of diesel produces 2.7 kg of carbon dioxide. Some of the effects of exhaust emissions are :

- Headaches
- Nausea
- Vomiting
- Coma
- Eye irritation
- Asthma

Lead poisoning and possibly even death of 99.3 per cent carbon dioxide is emitted by road transport only.

The main exhaust gases are:

Nitrogen (N_2)—This gas is originally in the air and takes no part in the combustion process but is sucked in through the vehicles air filter.

Carbon Dioxide (CO_2)—This gas is the product of complete combustion of the fuel.

Carbon Monoxide (CO)—This gas is the incomplete product of combustion.

Hydro Carbons (HC) — This is un-burnt or partially burnt fuel particles in vapourized form.

Water Vapour — Another product of complete combustion

Particulates (especially from diesel engines) — These are very small particles from un-burnt fuel. They may have hydrocarbons and other complex organic compounds absorbed on the surface. They are also known as Volatile Organic Compounds (VOCs).

Nitrous Oxide (N_2O)—This is another gas that can sometimes be produced.

Nature or Problems

The above said exhaust of gases because of the emission of smoke by various types of industries leads to face the following problems:

Avalanche
Melt of the Ice caps on the poles
Holes in the Ozone layer
Low rainfall or Non-rainfall even during rainy seasons
Global warming
Affects in Green productivity
Affects in pure drinking water
Affects in ground water levels
Some peculiar diseases like SARS, Small Pox, Blake, skin diseases, unknown unnamed diseases etc.
Endangered spices, animals, herbals, medicinal plants etc., and
Impotency among males

These are now-a-days quiet common only because of the environment pollution and the discussion is to manage these issues.

For Your Information

Europeans and North Americans (i.e., only 10 per cent of the world population) own about 60 per cent of the world's car. Think about the problem created by their sophisticated life and they are going to fix us into a great headache in near future in the environment pollution. At the same time they started to think of the alternatives to be used to save the earth from all such evils. Hence, the bio-diesel, electric power operated machines, car, avoiding of luxury items like car and started to use bicycles to increase the green productivity and as well as their own individual physical condition to be 6 fit to live in the world in the rest of the periods.

The only countries going to suffer a lot are developing nations like us and hence forced to talk on these challenges at this juncture. Now only we started to introduce all kinds of advanced machines in full sphere in order to achieve developments by competition the developed nations in the world. By means of such aim we are silently inviting the above said problems and as we are not having basic infrastructure to escape from these problems it is better to think well in advance by our effective and efficient managerial qualities.

What can we do? Voluntarily or by Compulsion of Law

To take the initiation it is better to quote the judgment of the Honourable High Court of Madras Bench on 20th December 2004. Honourable Justices stressed and the ordered the State Government to take intensive action against the pollution especially in the Capital of the state. They added in their order that nearly 20 per cent of the vehicles only obtained acceptable emission certificates and operating their vehicles on the other hand the remaining 80 per cent of the vehicle operators have not obtained or undergone such checking for their vehicles through a recommended emission checking points moreover they are operating their vehicles and causing the environment pollution. This should be taken seriously and all the vehicles have to be checked immediately in order to protect the Capital first.

The Capital of the state bagged with highly educated people and having lot of scope to be aware of such dangerous pollution but not taking utmost care to save our mother earth. The ultimate sufferers are nobody else except us. And we are so adamant to enter into action to increase the green productivity and hence the law enforcement people are always taking steps to save the earth and as well as the public as noble service.

We should not wait for the implementation of law but to the best of our knowledge we ourselves start functioning of indulging in such harmful activities which spoil our life. If you don't want to live at least leave others to live by adhering the environment issues and how to manage it.

Nature presents all good things to us to live peacefully, healthily, happily and it is our responsibility to save environment.

Now-a-days because of the population density forest lands, fertile lands are becoming residential places. New skyward constructions, buildings, occupied lot of cultivable lands. The forest areas slowly disappeared by way of occupation by human beings by means of migration for various reasons and hence leads to have the problem of endangered species, animals, medicinal herbals etc.

Because of the removal of forest areas trees are smashed and the loss of such natural resources leads to have non-rain conditions and monsoon failures further leads to loose employment opportunities in agriculture. And hence the people are thronging the cities for job opportunities.

At present, the developed nations have come to understand this serious problem and get ready to start their businesses and industries in developing nations in the name of globalization.

Due to the innovation in the field of science and technology we are force to introduce new machines for faster growth in the world level the small and medium enterprises are becoming endangered one. Hand made products say handicrafts, tiny industries, and small scale industries, cottage industries everyone of such is trailing and struggling to live. They slowly started to install and introduce high level machines in the name of modern technology for survival. Countries like India have failed to give much more importance for the green productivity. Industries are needed but the emissions from such industries should not kill us. Once we let the nature to live peacefully then it will allow us to live with peace if we fail to do so or disturb it will make us into so many pieces.

Will we allow our house with full of Garbage? *Then why in roads...*

We are throwing garbage on streets, smoking in public places, not checking our vehicles by proper maintenance regarding emission, using advanced electrical appliances which emit smoke harmful to the green productivity, not ready to dedicate ourselves in maintaining green house and so on.

When we are in Singapore, Malaysia even we hesitate to spit or spew out any mode of garbage on roads but while entering Indian Airport the first thing we are doing is the above said banned activities. It is because of the severe implementation of law in such foreign countries but our nation is blinking to implement sincerely for the welfare of their people to retain their vote bank.

Managing the Environmental Issues

Pollution control generally involves preventing the facility from harming people or the environment and mitigating the effects of any pollutant emissions. Harm to people or the environment is difficult to prove so regulations are increasingly designed to eliminate contamination. Effluent standards are based on health effects studies but because of factors such as cancer's long latency period, and the lack of human exposure data, standards generally include a large safety factor. The environmental manager must recognize that there is little margin for error when it comes to contamination issues.

Managing hazardous materials generally involves the following four major issues: identification (air and water emissions, rain waters, spent raw materials, etc.) storage and handling, disposal and shipping, and recordkeeping. An industrial facility that does not have programmes addressing all the four of these issues should essentially not be handling hazardous materials.

Good record keeping and communication are essential to several key elements of any pollution control and prevention programe: emergency procedures, contingency planning, and employee training.

Well run organizations are those which have established formal Safety and Hazardous Materials Management Programmes. The need for an integrated programme and policy uniting everything from purchasing through use and disposal can be demonstrated by considering what happens if a hazardous material is purchased or generated without knowledge of the regulations.

Disposal issues are increasingly important because federal regulations are dynamic, or changing. Waste minimization requirements and the land disposal ban are making waste production more expensive. There must be spill containment, supplies for cleanup, controlled access, and segregation of incompatible materials. Both on-the-job training and a written training programme are necessary. Fines may result if the training is not properly documented and employees are not tested for competency.

A company cannot control pollution or its hazardous wastes by selling contaminated properties or assets, however, it can certainly add to its liabilities by purchasing contaminated property. Unfortunately, many companies already own contaminated property with projected remedial actions that may cost millions of dollars. Too often, remedial action contracts permit a consultant to design and build with little or no oversight. If such contracts are not managed properly cost overruns and disputes can be expected. There must be a management plan that includes numerous inspections and thorough documentation. There must also be emergency plans in the event that something goes wrong such as a spill, a fire, or explosion.

Managing Toxic Substances

Toxic substances can create pervasive environmental and public health problems. The sheer volume of toxic materials manufactured and the many avenues of exposure (occupational, consumer use, and environmental residues), greatly increase the unacceptable health and environmental risks from many of these substances. Public policy has traditionally been aimed at protecting the public from toxic substances.

The Occupational Safety and Health Act is the primary federal law regulating toxic substances to protect workers in the workplace.

Table 16.1 : Areas of Concern Addressed by Federal Toxic Substances Statutes

Area of Concern	*Federal Statute*
Occupational Protection	• Occupational Safety and Health Act • Superfund Amendments and Reauthorization Act
Environmental Pollution	• Clean Air Act • Clean Water Act • Safe Drinking Water Act • Resource Conservation and Recovery Act
Chemical Manufacture and Use	• Federal Food, Drug, and Cosmetic Act • Federal Insecticide, Fungicide, and Rodenticide Act • Toxic Substance Control Act • Superfund Amendments and Reauthorization Act
Transportation	• Hazardous Materials Transportation Act • Resource Conservation and Recovery Act
Cleanup Actions	• Comprehensive Environmental Response Compensation, and Liability Act (as Amended by SARA)

The main provisions of the Act dealing with toxic substances include:

(a) Establishing and enforcing standards to limit exposure to various chemical substances that could induce acute or chronic health effects.
(b) Regulating substances that may cause cancer.
(c) Informing employees of the dangers posed by toxic substances through the use of Material Safety Data Sheets.
(d) Requiring employers to maintain medical, training, and other records to track the development and incidence of occupationally induced disease.

Conclusion

Slowly we are consuming poisoncus gas emitted by machines operated by men at work, vehicles using for transportation, industries using chemicals, using fertilizers for faster growth, using pesticides for protecting the green, using the leather garments, luxury items which emit harmful smoke, sale of tobacco products, and so on. There is no nature there will not be human beings.

The problems are in multi-dimensional sides but how to solve it is also in our hand. *Why don't we manage it?* Finding alternative like *bio-diesel, electric power, hydro-electric, wind solar, and wave, ethanol and tidal* are also emerged still **Why do we hesitate to introduce? -** *to save the mother earth ...*

REFERENCES

Aulay Mackenzie, Andy S. Ball and R. Virdee, *Instant Notes - Ecology.*

Nicholas P. Cheremisinoff and Madelyn L. Graffa, Publisher : *Environmental Health and Safety Management,* Noyes Publications/Jaico Publishing House, Mumbai, ISBN 81-7992-063-1: First Edition: 2003.

P.D Sharma, *Ecology and Environment,* Rastogi Publications - New Delhi

S.S. Purohit and R. Ranjan, *Ecology, Environment and Pollution*, Agrobios (India) Publications - Jodhpur.

V. K. Prabhakar, *Environmental Management,* Anmol Publications Pvt. Ltd., New Delhi.

17

ENVIRONMENTAL MANAGEMENT ISSUES IN URBAN AREAS

Srinivas Gumparthi

> *The reasonable man adapts himself to the World : the unreasonable one persists in trying to adapt the World to himself. Therefore all progress depends on the unreasonable man.*
>
> —*George Bernard Shaw*

India fortunately has a rich natural resource base including streams, lakes, and rivers, wet lands, forests, and good agricultural soils. Farmers and forest landowners provide a wealth of benefits from sound management of thses resources. However, some land management practices can result in relatively high environmental risks, whole others present low risk or virtually no risk at all. The natural resources on your land and farm can be protected by applying various practices that deal with the broad range of management choices available in today's agriculture and forestry. Water quality and quantity, soil quality, pesticide and nutrient use, wildlife habitat, forestry and waste management are all natural resources management issues that must be addressed as a part of the farm or forest operation. Ignoring these issues can affect the quality of the soil and water resource while risking public health as well as that of you and your family. This paper presents simple environmental management issues information about environmental management and pollution prevention.

Environmental Issues

Balancing responsible corporate behaviour against commercial imperatives is a constant challenge. Government of India and NGOs has already established clear environemental policies and continues to work on implementing these across its entire business and supply chain. The main objective is to reduce any negative impact which its activities might have on the environment.

Spreading Best Practice

The Government of India and Environmental Concern agencies are co-coordinating to implement environmental management issues and best practices through a dedicated corporate Environment Working Groups. The Government of India is trying to identify and spread best practices across the nation.

The focus should be on a number of Key Performance Indicators (KPIs) for environmental management, particularly:

- Reducing emissions of green house gases now and planning for further reductions.
- Assessing the environmental impact of the company's business activities, setting targets and producing reports.
- Greater energy and resource efficiency.
- Moving away from disposing of waste towards waste minimization, recycling and recovery.

Within each of these areas, there are specific targets and indicators are to be established to measure its performance.

Key Activities and Objectives

The environmental management system should include the following:

- Setting and monitoring KPIs for environmental managament.
- Implementing certified ISO-14001 environmental management systems across all business units.
- Introducing specialist environmental management tools, such as utilities management software.

- Promoting and recognizing good environmental management in the field, by rewarding local depots and offices for good practice.
- Moving increasingly towards renewable energy resources and reducing water and energy consumption.
- Implementing programmes to reduce waste and emissions
- Actively protecting endangered wildlife and environmentally sites.
- Encouraging better environmental management as part of education system.

Social Responsibility

We do not inherit the Earth from our Ancestors, ... we borrow it from our Children

—Antoine de Saint-Exupery

With over one billion people in India everyone's responsibility to the society has never been greater. Altogether, over 80 per cent of the people who live and make their life by drawing environmental resources. Managing this environmental system is underlined by its self-commitment of the people in meeting the mission afloat is very important.

Providing a Safe and Supportive Working Environment

Individuals and teams across the country often work in pressured, difficult and potentially dangerous situations. So every member of the society should take up the responsibility of protecting environment and should provide services in keeping up the environment clean and green throughout. This is the lifetime challenge of every human being to safeguard the environment.

Key Activities

- Developing, implementing and regularly updating fair and flexible human resources policies.

- Encouraging a common corporate culture, based around responsible as well as profitable business practices on sound environmental practices.
- Working with NGOs to establish local and national agreements on relevant environmental management areas.

Community Involvement

Although each one of us faces a wide spectrum of social and community challenges day in and day out. Every individual should own the accountability this will be big positive environmental driver. Even our business activities should contribute to the well-being of society as a whole, and its members should retain the strong sense of the public service ethos which comes from providing front-line services to the community.

> *"I feel more confident than ever that the power to save the planet rests with Individual Consumer."*
>
> —*Dennis Hays*

Accountable to Society

Every business has an impact on society. Every business should recognize its role and duties, not just as an employer in the communities where it operates, but also as a key contributor to the development and regeneration of these areas. This investment, both financial and non-financial, not only helps the community but underpins the company's ability to grow in the future. Business is about people and community. The companies should manage their resources to provide a safe, secure and pleasant environment in which people can live, work, learn and travel.

Support for Local Communities

Key Activities

- Promoting community consultation and involvement in

environmental projects and business activities, as an integral part of the community relationships.

- Understanding the social impact of the business activities on local communities.
- Retaining a strong public service ethic amongst the people, and creating an appropriate balance between social responsibility and business profitability.
- Encouraging local involvement in community, social and environmental initiatives and projects.

Urban Environmental Management

Take any of today's environmental problems faced by the inhabitants of Earth, and its causes and pressures can easily be traced back, directly or indirectly, to urban areas. The forces and processes that constitute 'urban activity' have far-reaching and long-term effects not only on its immediate boundaries, but also on the entire region in which it is positioned.

In a very broad sense, the urban environment consists of *resources*, human and other; *processes*, that convert these resources into various other useable products and services; and effects of these processes, which may be negative or positive.

Resources	*Processes*	*Effects*
• Human Resources	• Manufacture	• *Negative Effects*—Pollution-air, water, noise—*Waste,* Generation—garbage, sewage Congestion, overcrowding
• Sunlight	• Transportation	
• Land	• Construction	
• Water	• Migration	
• Minerals	• Population Growth	• *Positive Effects*—Products, Value-addition, Increased knowledgebase/education, Access to better services
• Electricity	• Residence/Living	
• Fuels	• Community Services	
• Finance	(Education, Health ....)	
• Intermediary products		
• Recyclable materials		

With the inevitable danger of overlap and generalization, the following three dimensions are identified as core aspects in urban environments:

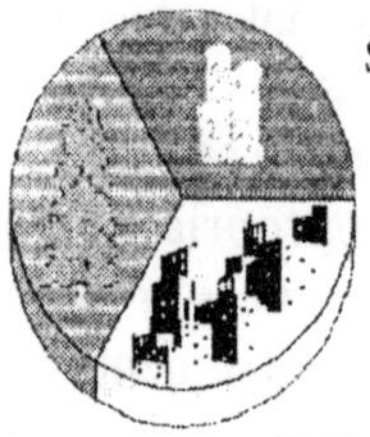

Natural Environments

Resources, processes and effects related to flora and fauna, human beings, minerals, water, land, air, etc.

Built Environments

Resources, processes and effects related to buildings, housing, roads, railways, electricity, water supply, gas etc.

Socio-economic Environments

Resources, processes and effects related to human activities, education, health, arts and culture, economic and business activities, heritage—urban lifestyles in general.

It is the intersection and overlay of these three dimensions that constitutes an 'urban environement'. Taking anyone dimension at the exclusion of the other two poses the inevitable danger of missing the forest for the trees—the interdependency and interdisciplinarity of the three dimensions have to be fully understood in the development of coherent and sustainable policies and programmes for the urban environment.

This is particularly true with the multiplicity of actors and activities—there has been a growing realization that state agencies and activities are, but one part of a spectrum of agencies and activities that are involved in the urban environment. These and other issues are covered in depth in this document.

In order to develop an effective response to the myriad range of challenges facing urban environments today, the following framework has been developed to tackle these and related issues in the formulation of policy and practice of urban environmental management.

The basic aim of this framework is three-fold:

- To develop awareness and educate on issues related to urban environments;
- To assist in policy and programme development; and
- To facilitate monitoring and evaluation.

The target audience of this framework should be kept broad to increase its utility value—government agencies, planners and planning bodies, NGOs, donor agencies, community groups, academics etc.

The framework is presented as a series of statements. Each statement is briefly explained and is supported by resources available.

Urban Environments

Urban Environments pose a challenge for effective distribution and management of global resources

As illustrated in the Numbers section, the density and population of today's cities necessitates the equitable distribution of resources that are needed for its various activities. It is, therefore, necessary to understand the effects of an urban area not only within its immediate boundaries, but also on the region and country it is positioned, due to the large amount of resources necessary to sustain it.

Resources addressing this issue:

- *Tokyo's Footprint*

Balancing Acts

There is a need to strike a balance between natural and built environments and between ecological and economic objectives.

Agglomeration and centrality of resources and skills that an urban area offers should not be ignored, but should be balanced with the natural environment, including air, water, land, and mineral resources. Economic objectives in job creation, income generation and distribution, particularly for developing countries, will have to be tempered with ecological objectives of sustainable living. The priority that developing cities place on economic development and income distribution over that of environmental issues has to be understood from the larger perspective of long-term human development.

Resources addressing this issue:

- *Urban Share of GDP*
- *Green Consumerism*

Visioning

There is a need to develop a structure of goals/visions and a methodology to achieve it in order to identify the action that has to be taken

A structure of goals and visions for sustainable urban living that can easily be understood by ordinary citizens should be developed. This will allow communities and Governments to discuss ways in which this can be achieved. It also attributes legitimacy and currency for the problems faced in urban environmental management, and sets the platformon which these problems can be addressed. The scale of Urban Environmental Management problems should be understood, so that appropriate action can be taken at the appropriate level.

Resources addressing this issue:

- *Environmental Impact Assessment*

Action plans

Steps need to be taken that are relevant in the short-term in order to gain wider acceptability, keeping long-term goals in mind.

In relation to the previous point, the goals and visions have to be divided into immediate, intermediate and eventual goals, so that the issues are better understood, and tangible/visible results can be achieved. Sharing and cooperating on essential lessons, practices and technologies is critical to achieving such goals.

Resources addressing this issue:

- *Scale of urban problems*
- *Urban Governance*
- *Capacity Building*

Information Management

Access, sharing and dissemination of information has to be a priority to achieve greater understanding of the issues involved.

The cause-and-effect reasoning of local actions has to be understood from a regional and global perspective. For example, what is the effect of smoking in public places? Key to achieving this understanding and exploration is information, and its easy, adequate and immediate access. Timely and packaged Information is key to influences local decision-making processes, which on a cumulative basis, have global repercussions.

Resources addressing this issue:

- *Information Management Issues*
- *Information and Urban Environments*

Knowledge Transfer

Collaborative efforts in 'knowledge transfer' at the city-to-city level has to be encouraged, particularly between developed and developing cities.

Collaboration with institutions and governments in developed and developing countries for the transfer of urban 'software' (Inspiring ideas and best practices, innovative technologies, practical solutions, etc.) has to be encouraged-covering among other issues, policies, programmes, skills, local and city governance. Feasibility and transferability of such software will have to be studied in depth before collaborative projects are launched.

Resources addressing this issue:

- *Insights*
- *Programmes and Initiatives*

Sustainability

There is a need to understand and enact the concept of sustainable development and sustainable living, in all its varied definitions.

Wider participation to achieve the goals of sustainable development and living has to be encouraged. This has to involve the community, local governments, and the whole range of non-governmental organizations (including the private sector). Sustainable living should become a way of life, rather than a fancy concept that is espoused by a 'enlightened' few.

Resources addressing this issue:

- *Environmental Management Tools*
- *Sustainable Communities and Participation*
- *Green Consumerism*

Technology

Development of new technologies that are clean, green, and practical has to be encouraged and exchanged between local and city governments to combat environmental problems.

Consequences of current polluting technologies has to be enabled through a variety of governmental and non-governmental forums, including those online.

Resources addressing this issue:

- *Notes Technology Transfer*
- *Voluntary Initiatives to Mitigate Climate Change*
- *Green Business Practices*

Ten Principles of the New Environmentalism

There has been a growing change in the way environmental policies, programmes and projects are formulated and implemented. Among the key features of this 'new thinking' has been the linking between economic activities and environmental problems; the need for a multiplicity of actors and actions at various levels to solve these problems; the cross-border, global nature of environmental problems needing actions to be taken at the international level.

Specifically, the new environmentalism—characterized by greater precision in factoring environmental costs and benefits in to policy making—puts local people at the centre of environmental strategies, diagnoses and addresses behavioural causes of environmental damage, and recognizes the political dimensions of environmental reform.

Ten principles characterize the new environmentalism in force today:

Principle 1

Set priorities carefully: The seriousness of environmental problems and the scarcity of financial resources have required tough prioritization and phasing of remedial actions. Broad, shallow and expensive approaches have to be avoided, with serious priority-setting exercises grounded in proven studies and analyses.

Principle 2

Make every rupee count : Unnecessarily expensive environmental policies, based on high-cost approaches traditionally used in industrial countries, have to avoided. There is a need for a new emphasis on cost-effectiveness, allowing for much more to be achieved with limited resources. It requires a multidisciplinary approach—one that calls for environmental specialists and economists to work together to identify the lowest-cost methods of addressing key environmental problems.

Principle 3

Harness "win-win" opportunities : Some gains in the environment will involve costs and trade-offs. Others can be achieved as by-products of policies designed to improve efficiency and reduce poverty. Reducing subsidies on the use of natural resources, clarifying and reallocating property rights, are examples of 'win-win' policies, which are expected to have beneficial impacts on the environment, particularly given the scarcity of resources devoted to solving environmental problems.

Principle 4

Use market instruments where feasible : Market-based incentives to reduce environmental damage are best in principle and often in practice as well. They contrast with traditional command-and-control and technology-driven regulations that have been the norm until recently. Innovative approaches involving emissions and effluent charges, harvesting permits market-based extraction charges and tradable permits are some examples.

Principle 5

Economize on administrative and regulatory capacity : In many countries, administrative and enforcement capacity is usually weak. They recognize that they cannot adopt highly 'enforcement-intensive' approaches and therefore emphasize on self-enforcing policies and other instruments (such as taxes, bans, fees etc.) with less intervention. This approach has also provided increased roles for NGOs and community groups.

Principle 6

Work with the private sector, not against it : Recognizing their limited regulatory capacity and the need for accelerated private investment, many Governments are switching from a control-dominated attitude towards the private investment sector to one that involves dialogue and negotiated, monitorable programmes.

Self-enforcement, independent certifications schemes (for example the ISO-14000) are now playing a much larger role.

Principle 7

Involve citizens throughly : When a country's environmental problems are adddressed, the chances of success are greatly enhanced if local citizens are involved. Such involvement is needed for four reasons:

1. Local citizens are often better able than governments to identify the priorities for action.
2. Members of local communities often know about cost-effective solutions that are not available to governments.
3. The motivation and commitment of cummunities are often what sees an environmental project through its completion.
4. It helps build constituencies for change as a counterweight to vested interests.

Principle 8

Invest in partnerships that work : Often, the most effective way in dealing with environmental issues is when the stakeholders work in partnership. NGO involvement in priority-setting exercises, and tripartite relationships, (the government, the private sector and the community organizations) are becoming increasingly common. The value of such partnerships stems from not only the different perspectives and skills that are brought to the table, but also the necessity of carrying out concerted actions to address environmental issues.

Principle 9

Remember that management is more important than technology: The old-fashioned, technology driven approach to the environment is giving way to recognition of the crucial role of good management. Improved management practices are always a complement to, and sometimes a substitute for, investment in equipment.

Principle 10

Incorporate the environment from the start : When it comes to protecting the environment, prevention is much cheaper—and more effective—than cure. Most countries now seek to assess and mitigate potential damage from new infrastructural investment. Efforts to move such concerns 'upstream' in the project cycle to factor environmental concerns into sectoral strategies, are underway.

"Even if I were certain that the World would end tomorrow,
I would plant a tree this very day."
—Martin Luther King Jr.

REFERENCES

Agarwal, Anil, "Human-nature interactions in a Third World country", Fifth World Conservation Lecture, The Environmentalist, vol. 6,No. 3.

Agarwal and Narain, "India: The State of the Environment," 1985.

Annual Administrative Report for the Tehri Garhwal State 1937.

Adakari, Ajay. Interview. 13 February 1995. Professor of Accounting at The American University.

Bahuguna, Sunderlal, "*Uttrakakhand Mein Ek Sau Bis Din,*" in Hindi, Dehradun. 1974.

Business International. "India Goes Greener in 1990." 17 June 1990.

Gross, Birth Death and Migration in the Himalayas, New Delhi: p. 184.

Guha, Ramachandra. "The Malign Encounter: The Chipko Movement and Competing Visions of Nature." United Nations University/World Institute for Development Economics.

Guha, Ramachandra.*The Unquiet Woods.*University of California Press. Berkeley. 1989.

Guha, Ramachandra. "Ecological Roots of Development Crisis" EPW, 12 April 1986.

Inter Press Service, "Modern-day Gandhi Forces India to Suspend Dam Project" 10 June 1992.

Joshi, Gopa "Men propose women dispose" *Indian Express*, 14 January 1982.

Meadows, Donnella, *et al.* The Limits of Growth. New York, 1971.

New York Times. "Social Activism Sprouts as India's Politics Decay", 15 June 1982.

Reuter News Service India, "India Green says Dam Hurts Hindu Feelings", August 25, 1994.

Smythies, E.A. "The resin industry in Kumaun, Forest Bulletin No. 26 Calcutta. 1914.

18

ENVIRONMENAL MANAGEMENT FOR SUSTAINABLE BUSINESS

V. Pugazhenthi and *C. Sunitha*

There is a common criticism that companies seek business opportunities in everything including in the environmental disasters created by their own factories, technologies and products. Even the "Green Business Summit 2003" held at Hydrabad was termed as an effort in green washing corporate crimes, and the corporate are shedding crocodile tears. At the juncture of Union Carbide and Dow Chemicals, shirking their responsibility towards the Bhopal victims, this kind of criticisms is inevitable.

The establishment of voluntary environmental standards is providing impetus for companies to adopt worldwide best practices, which emphasize a more integrated approach to environmental management. Some such standards include international standards organization's ISO-14,000 the International Chamber of Commerce's (ICC) Business Charter on Sustainable Development, the Global Environmental Management Initiative's (GEMI) Environmental Self-Assessment Programme, and Total Quality Environmental Management (TQEM) Programmes.

Corporate Environmental Management

Companies are beginning to realize that environmental issues to be addressed for a number of reasons. Including: Consumer pressure, potential cost savings, legislation and ethics. There is, therefore, growing interest in the area of corporate environmetal management. Firstly, they must recognize the strategic advantage, which corporate environmental management can bring and to ensure that commitment to environmental improvement exist in the business.

Tools of Corporate Environmental Management

1. Environmental Management Systems—the organizational structure, responsibilities, practices, procedures, processes, and resources for determining and implementing environmental policies.
2. The Standardization of Environmental Management Systems—complying with international standards like ISO-14000.
3. Environmental Policies—the overall environmental intentions and directions of an organization regarding the environment.
4. Environmental Guidelines and Charters—documents produced by NGOs with the general aim of helping businesses manage and improve their environmental performace.
5. Lifecycle Assessment—An analysis covering every stage and every significant environmental impact of a product from the extraction and use of raw materials through to the eventual disposal of the components of the product and their decompositon back to the elements.
6. Measuring Environmental Performance—including qualitative, quantitative, contributor and external measures.
7. Environmental Reporting—Various means by companies disclose information on their environmental activities.

Strategies of Environmental Managememt

As it is summarized by Richard Welford, the companies can be classified into ten categories based on their invelovement in CEM, as follows :

No.	*Strategy*	*Description*
1.	Ostrich	Companies that assume environmental challenge is a passing fad
2.	Resistant	Companies that hinder the passing of environmental laws and regulations.
3.	Publicity	Companies in which some well publicized event or accident acts as a catalyst.
4.	Indifference/non-compliance/stable/passive/laggards/ignored	Companies having low environmental risks, low enviromental returns, cost constraints etc.
5.	Thinker	Companies waiting for others to take the lead.
6.	Offensive/smart movers	Companies having high environmental returns
7.	Defensive/compliance/reactive/localized action	Companies having high environmental risks
8.	Compliance-plus/anticipatory/doers/corporate (proactive)	Companies that move beyond compliance
9.	Commercial and environmental excellence/entrepreneurial	Companies where there is clean technology and organizational refo m
10.	Innovative/enthusiasts/leading edge business scope action	Companies having high environmental risk, and also high return

Source : Richard Wellford, 1996.

Pressure to adopt CEM

Following are the pressures that are exercised upon the corporate to implement an acceptable environment management system (www.epa.gov.).

1. Government Pressure

Perhaps the most obvious stakeholders that influence firms' adoption of environmental practices are various government bodies. Legislation authorizes agencies to promulgate and enforce regulations, a form of coercive power.

In particular, Delmas (2002) found that governments play an important role in firms' decision to adopt ISO-14001. First, governments can act as a coercive force by sending a clear signal of their endorsement of ISO-14001 by, for example, enhancing the reputation of adopters. Second, government can facilitate adoption by reducing information and search costs linked to the adoption of the standard by providing technical assistance to potential adopters. Political pressure is the level of political support for broader or more stringent regulations. Regulatory pressure represents the extent to which regulators threaten to or actually impede a company's operations based on their environmental performance.

2. Customer and Competitive Pressures

In addition to government actors, firms may facilitate coercive and mimetic isomorphism. For example, multinationals are widely recognized as key agents in the diffusion of practices across national borders by transmitting organizational techniques to subsidiaries and other organizations in the host country (Arias & Guillen, 1998). Firms may also mimic practices that successful leading firms have adopted. In addition, firms respond to customer requirements. The customer supplier relationship is perhaps the primary mechanism through which quality management standards have diffused.

Several studies have found that firms that adopted environmental management practices were motivated by customer concerns. A survey of the largest Canadian firms showed that customer pressure was the second most cited source of pressure to adopt an environmental management plan, after government pressure (Henriques & Sadorkey, 1996). Khanna and Anton (2002) found the consumers exert more pressure on companies to adopt environmental management practices than commercial and industrial customers. Christmann and Taylor (2001) showed that customers in developed countries have influenced companies in China to improve their environmental compliance and adopt the ISO-14001 EMS standard.

3. Community and Environmental Interest Group Pressures

Communities can also impose coercive pressure on companies

through their vote in local and national elections, via environmental activism within environmental non-government organizations (NGOs) and by filling citizen lawsuits. Several studies have found that company decisions to adopt environmental management practices are influenced by the desire to improve or maintain relations with their communities.

Another study based on a survey of ISO-14001 certified companies across 15 countries found that one of the strongest motivation factors to pursue certification was the desire to be a good neighbour (Raines, 2002). There are many examples where companies have amended their environmental practices in response to environmental group pressures (Baron, 2003). For instance, after Mitsubishi Corporation was subject to a protracted consumer boycott led by the Rainforest Action Network (RAN), Mitsubishi announced it worked no longer use old-growth forest product (World Rain-forest Movement, 1998).

4. Industry Pressure

Market concentration within an industry may also affect the rate of diffusion of environmental management paractices. If a few big players dominate an industry it may require their suppliers to adopt particular CEM Parctices. If this is likely to lead to a greater diffusion of these practices than if the industry were more fragmented. This is a major reason why automotive suppliers in the United States have adopted similar quality and environmental practices.

Interactions

The interaction between these institutional pressures is likely to moderate their individual influence on company practices. For example, the pressure from environmental groups may encourage the formulation of more stringent regulations. This, in turn, can induce industry leaders to encourage laggard firms to adopt environmental pactices. Similarly, following its chemical disaster in Bhopal in 1984, Union Carbide along with other large chemical companies faced mounting public pressure for more stringent safety and environmental regulations. In response, the chemical industry developed and promoted a set of Environment, Health and Safety (EHS) management practices—the Responsible care programme—to chemical industry associations in Canada and the United States (King & Lenox, 2000).

Environmental Management System (EMS) in Developed World

Majority of the Multinational Enterprises (MNEs) in the OECD countries are following the management tools such as the ISO-14000 series, AA-1000 and OECD Guidelines. The background report for the 2004 roundtable on corporate responsibility points out that apparently high share of companies in the industrialized economic publishes environmental policy statements (EPS). The EPS is a statement by an organization of its intentions and principles in relation to its overall environmental performance. As per a sample study by OECD, by September 2003, 58 per cent of the companies had issued statements that meet certain minimum requirements with reference to key issues including suppliers, contractors, resources, and materials, energy use and efficiency, emissions to water, emissions to air, transport, waste minimization/reduction/ disposal and recycling, packaging, product and /or stewardship/ design, social impact, noise, neighbourly concerns, visual blight, employee training, greenhouse keeping sustainability and industry specific issues.

This study reveals that in developed countries like United Kingdom the companies completely comply with the relevant laws. Similarly, nearly 20 per cent of the companies are found involved in any of the four voluntary initiatives. Namely the International Chamber of Commerce's (ICC) Business charter for sustainable development, the Coalition for Environmentally Responsible Economies (CEREs), Corporate Environmental Reporting Requirements, UNEP's Finance initiative and chemical industry's responsible care.

Another OECD study, surveying a total of 4,176 enterprises, concludes that there is a statistically positive significant relationship between company size, whether the company is part of an association, profitability, whether the market is international, whether the head office is overseas, the presence of a quality management system and whether regulatory incentives were provided to.

EMS in Small and Medium Enterprises

Several studies across the world reveal that small and medium enterprises place the environmental management low on their list of priorities, because of the following reasons :

- SMEs have limited managerial and financial resources.
- SMEs may be inhibited by the perceived cost of CEM.
- SMEs may consider themselves to be in compliance with the international environmental standards.
- SMEs may feel the pollution control measures taken and equipments, they have are adequate.
- SMEs may think that its environmental responsibility is over, if they entrust their waste disposal programme to a professional agency.
- SMEs may have few major customers in a closed supply circuit.

So, to recognize the importance of environmental management and to adopt the same, the SMEs are to be aware of :

1. Legal and regulatory requirements from polluiton control authorities.
2. The environmental aspects of their activities.
3. The direct, indirect and financial cost and benefits associated with these environmental aspects.
4. Growing green purchasing behaviour on the part of consumers.
5. What competitors achieve or implement in environmental management.

So, SMEs can no longer take the "ostrich" Strategy assuming environmental challenge is a passing cloud and they should realize the potential drawbacks of not introducing environmental management strategies such as market exclusion, customer damage claims, reduction of sales, increased insurance costs, possible legal costs and fines.

SMEs and Eco-Industrial Parks in India

An eco-industrial park is a community of manufacturing and service businesses seeking enhanced environmental and economic performance through collaboration in managing environmental and resource issues including energy water and materials. In India, in addition to the initiatives taken by the Ministry of Environment and Forests, Government of India, several agencies have taken initiatives to propagate the concepts of cleaner production. With assistance from the National Cleaner Production Centre, state cleaner production centres have been set up in Gujarat, Karnataka, Punjab, and West Bengal (Rathi 2003).

Most Industrial Estates in India comprise a large number of SMEs, which lack personnel and financial resources to implement efficient EMS. The SMEs also produce a large variety of wastes but in relatively low amounts. (Resources, Energy and Development 1: 87.) Economic Development in developing countries like India can take place through well-planned IEs where many SMEs are clustered together regionally. This regional agglomeration of SMEs in IEs provides a good opportunity for eco-industrial networks (Hanf and Wilderer, 1999).

Wimby Syndromes of SMEs

In an attempt to clean their own environment, the industrialized countries export the hazardous wastes and Domestically Prohibited Goods (DPGs the products that are either banned or severely restricted for sale or consumption in the country of origin but are freely allowed to be exported) to other countries. Each day, the world produces more than a million tons of hazardous wastes. According to UNEP's estimates, approximately 338 million tons of hazardous waste is generated annually. The majority of this waste is generated in developed countries nearly 10 million tons of toxic garbage cross international border every year.

India imports hazardous wastes from many countries including Australia, Canada, and the US & UK. Approximately 22,03,182 kg of metal wastes were shipped to India during 1989 from Germany and 5,44,773 kg of plastic waste from the US in January 1993. According to Green Peace International Toxic Campaign in 1995 alone, US exported about 45000 metric tonnes hazardous waste consisting of lead, zinc ash, dross, residue, slag and scrap besides plastic, copper, aluminium. While in just 2 months in the same year (July and Ausust), Australia shipped over 2000 mt of such waste to India. In the 4-month period between December 1995 and March 1996, about 1533 metric tonnes of lead waste was imported into India (*www.ilo.org*). Most of the waste thus, exported to developing countries is considered too toxic to be recycled or dumped in the exporting country.

To recover and reuse the metals at cheaper price in hazardous waste, the developing world goes for hazardous wastes shipments from industrialized countries. The Basal Convention on Tran Boundary Movements of Hazardous Wastes and their Disposal allows shipments only if the receiving country can manage the waste appropriately. Still, the US sends lead acid batteries to Asian countries under the guise of these being recoverable or recyclable waste. The lead may be recovered but acids and other battery contaminators are dumped in the environment. The small and medium size enterprises often fall pery to this WIMBY (Welcome in My Backyard) Syndrome and stand to lose their business in the long run. The SMEs should comply with the rules and regulation in the country while importing and processing the waste, the breach of which may hurt not only their concern but also the immediate environment.

Sustainable Business in India

Nearly 36 per cent of all accidents during the twentieth century occurred in developing countries, especially those in Latin America and Asia. Similarly, if one looks back the history of industrial disasters that occurred in the second half of the last century, in terms of lives last, the Bhopal disaster is placed No.1. So, in India the CEM is the felt need today. Here are few examples for sustainable business practices followed in India.

- In India Hero Honda Motors Ltd. has geared itself to become a green Company by the end of the current fiscal year, as it is planning to produce its entire range of vehicles conforming to the year 2000 emission norms.
- Unilever is now publishing a separate environmental assessment every two years and has launched three major initiatives to conserve global fisheries, promote clean water and shift to 'sustainable agriculture.'
- The IPCL, Gujarat Refinery, GSFC, Namdesari industrial Estate and small industrialists are jointly setting up a 56 km long effluent channel in Vadodara.
- Bajaj Auto Ltd. has obtained ISO-14001 certification for strictly adhering to environmental norms in the manufacturing process.
- The ABB Company had planted 2,500 trees in its 100 acre premises as part of a green campaign.
- Responding to complaints from locals and naval officers in the region on the sulphur dust pollution, the Chennai based Murugappa owned complex fertilizers manufactures sent out a delegation of senior officers to Japan and South Korea scouting for pollution free ways to handle sulphur and finally, installed a sulphur handling facility for Rs. 10.5 crore.
- Some of the milk cooperatives are selling packaged milk with the offer of cash refund for the returned disposable plastic pouches.
- Henko and Fena, two detergent manufactrurers that have shown leadership have vision shifted to Zeolotes—a cleaner option to sodium tri-polyphosphate.
- Corporate Roundtable on development of strategies for the environment, a government body, has brought together

16 leading corporate including Ashok Leyland, BHEL, IOC, ONGC, TELCO, Reliance, ICI etc. to incorporate green products.

- Hand woven and chemical free fabric, herbal shampoo (Shanaz, Ayur) ayurvedic products, are examples of initiation of green movements by Indian companies.

Though there are instances of green movements in Indian Business as found above, systematic corporate environment management is yet to get recognized and incorporated by the industries.

Conclusion

It is sure that firms making and marketing goods with environmental characteristics will have competitive advantages over firms marketing non-environmentally responsible alternatives. But, in some cases firms have misled consumers as eco-friendly, in an attempt to gain market share. In other cases firms have jumped on the green bandwagon without considering the accuracy of their behaviour, their claims, or the effectiveness of their products. This lack of consideration of the true "greeness" of activities may result in firms making false or misleading "green claims". But the companies will definitely recognize in the future CEM as unavoidable and profitable management system.

REFERENCES

Agarwal, S.K. & Dubey, P.S., *Environmental Controversies,* API & Publishing Corporation, 2002.

Hanf and Wilderer, 1999, *Green Business Opportunitics* 5(2): 14-15.

Kattiyar, U.S., *Environmental concerns : Depleting Resources and Sustainable Development*, Pointer Publishers, Jaipur, 1997.

Sandeep Singh,"DPGs, Trade in Toxic Waste, and Technology Transfer: Issues and Developments", SAWTEE & CUTS-CITES, 2001.

Sinha, P.C., *Sustainable Development,* Anmol Publications Pvt., Ltd., New Delhi, 1998.

Richards Welford, *Corporate Environmental Management*, University Press, Hyderabad", 1999.

Rathi, A.K.A. 2003, *Journal of Cleaner Production* 11 (5): 583-590.
Unnikrishnan, Naik and Deshmukh 2004, *Energy and Development*, 1: 75-90.

Websites

1. http://www.cuts.org/Report-DPG.heml
2. http://www.ban.org/ban-news/ban-news.heml
3. http://www.ilo.org/public/english/dialogue/actracv/indene.html
4. www.greenbusiness.net
5. www.usaid.gov.in
6. www.pre.nl
7. www.coopamerica.org/cabn/conference/highlights.hetml
8. www,epa.gov/tri

19

IDENTIFICATION OF ENVIRONMENTAL ASPECTS AND SIGNIFICANT IMPACTS

B. Rajasekaran and *B. Nirmala Devi*

1. Identification of Environmental Aspects and Significant Impacts

Organisation of all kinds are in need of maintaining sound environmental performance by controlling the impact of their activities, products or services on the environment, by taking into account their environmental policy and objectives. They have to do so in the context of increasingly stringent legislation, the development of economic policies and other measures to foster environmental protection ultimately this may raise to the growth of environmental matters including sustainable development in the minds of interested parties like suppliers, customers etc.

Now-a-days many large scale and medium scale organizations have undertaken environmental "reviews" or "audits" to their environmental performance. But these "reviews" or "audits" may not be sufficient to provide an organization with the assurance that its performance not only meets, but will continue to meet, its legal and policy requirements. To be effective, "reviews" or "audits" need to conducted within a structured management system and integrated with over all management activity.

The major significant environmental aspects associated with the activities of industries are (a) contamination of land, (b) emissions to air, (c) releases to water, (d) waste management, (e) use of raw materials and natural resources and other local environmental and community issuses.

Government has enacted pollution control regulations to mitigate these problems but regulatory attempts in arresting environmental degradation have not been very successful.

In this critical situation, an approach like implementation of ISO-14001 Standards drives for the improved environmental performance.

In short to say, to improve environmental management, organization have need to focus not only on how things happen but also why they happening.

2. How to go about Creating and Environment Management System (EMS) in the Organization

The corporate sector should make itself accountable for achieving the environmental objectives of national and international framework. It should undertake environmental impact assessments for all proposed organisational plans, programmes and policies and integrate all of them into corporate decision making.

Paying adequate and timely attention for any environmental damages by corporate sectors comply them whole heartedly with national and international standards pertaining to environment. For creating an EMS a corporate sector may look into following steps:

Steps for Creating EMS

Commitment of Top Management

- Setting multi-disciplinary committee from Top Management
- Prepare initial Environmental Review report regarding current position
- Prepare a report covering requirements of ISO-14000, EMS registration and related cost benefits to the organization
- Prepare Top Management Commitee to the EMS process

Environmental Policy

- Identify actions to be taken to meet the requirements by the committee
- Proposal to meet ISO-14000 requirements by committee and top management
- Preparation of Environmental Policy by committee

Planning

- Conducting a review of organization operation by Top Management
- Identifying legal requirements and environmental concern
- Establish objectives
- Setting targets, formulates plan for meeting those targets

Implementation

- Implement the plan by establishing responsibilities, training and communication
- Maintaining documents
- Operating control procedures and audit procedures

Evaluation

- Perform final gap analysis, plan and find deviation
- Implement corrective action

Review

- Conduct review
- Check EMS is modified to optimize its effectiveness
- Prepare the organization for registration

Apply

- Apply for registration by outside auditors

3. Responsibilities of Hierarchy while Creating EMS

There is an increase in need for organization of all types to respond to the growing environmental concerns Top management has a key role to play in building awareness about environmental issuses to their employees and their commitment towards environmental policy. Because all members of the organization should understands and be encouraged to accept the importance of achieving the environmental objectives and targets for which they are responsible or accountable. For creating EMS, functions at various levels in hierachy are likely to be as under :

Top Level

Environmental Policies Set by BOD

- Identification of environmental aspects and significant impacts
- Identification of legal and other requirements
- Setting environmental goals, objectives, and targets that support the organisational policy

Environmental Policies Supported by Top Management

- Setting an environmental management programme
- Defining roles, responsibilities and authorities
- Setting programme of training and awareness procedures
- Explaining process for communication of EMS to the whole organization

Operational Level Practices

- Creating documents of operational and control procedures
- Setting procedures for emergency response
- Recording management procedures
- Procedures for management review

Departmental Level Practices

- Setting procedures for monitoring and measuring operation that can have a singnificant impact on the environment
- Procedures to correct non-conformance
- Programme for auditing and corrective action

4. Identification of Environmental Aspects and Evaluation of Associated Environmental Impact

An organizations' policy, objectives and targets should be based on knowledge about the environmental aspects and significant environmental impacts associated with its activities, products (or) services. An environmental aspect refers to an element of an organizations activity, product or service which can have a beneficial or adverse impact on the environment. An environmental impact refers to the change which takes place in the environment as a result of the aspect. The identification of the environmental aspects is an on-going process that determines the past, current and potential impact (positive or negative) of organization activities on the environment. The assessment of environmental loads related to a product/process/activity/services by identifying the aspects of quantifying the energy and materials used and their impacts of waste released into the environment is as:

Identification of Environmental Aspects and Evaluation of Associated Environmental Impacts

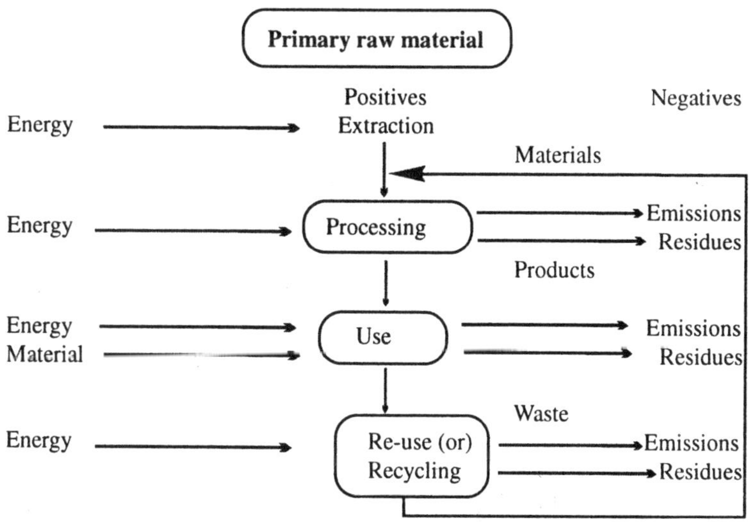

Assessment of Environmental Aspects and Evaluation of its Impacts

- What are the environmental aspects of the organization's activities, products and services?
- Do the Organization's activities, products (or) services create any significant adverse environmental impacts?
- Does the organization have a procedure for evaluating the environmental impacts of new projects?
- Does the location of the organization require special environmental consideration?
- How will any intended changes (or) additions to activities, products (or) services affect the environmental aspects and their associated impacts?
- How significant or severe are the potential environmental impacts should a process failure occur?
- How frequently will the situation arise that could lead to the impact?
- What are the significant environmental aspects considering impacts, likelihood, severity and frequency?
- Are the significant environmental impacts are local, regional (or) global in scope?

Phases Involved in Identification of Environmental Aspects and Evaluation of Environmental Impacts

Four major phases that involved for identification of environmental aspects and the evaluation of associated environmental impacts are likely to be as:

Phase 1. Select an Activity, a Product or Service

- The selected activity, product or service should be large enough for meaningful examination and small enough to sufficiently understood.

Phase 2. Identify Environmantal Aspects of the Activity, Product or Service

- Identify as many environemental aspects as possible associated with the selected activity, product or service

Phase 3. Identify Environmental Impacts

- Identify as many actual and potential, positive and negative, environmental impacts possible associated with each identified aspect :

Activity/Product/Service	*Benefit Aspect*	*Adverse Effect Impact*
Activity: Handling of hazardous material	Potential for accidental spillage	Contamination of soil or water
Product: Product refinement	Formulation of the product to reduce its volume	Conservation of natural resources
Service: Vehicle maintenance	Exhaust emissions	Reduction of air emissions

Phase 4. Evaluate Significance of Impacts

- The significance of each of the identified environmental impacts can be done with environmental concern and business concern even though it may differ for each organization.
- Evaluation of environmental concern may be facilited by the scale of the impact, the severity of the impact, probability of occurrence and duration of impact.
- Evaluation of business concern may be facilited by potential regulatory and legal exposure, difficulty of changing the impact, cost of changing the impact, effect of change on other activities and processes, concerns of interested parties and effect on the public image of the organization.

5. Lifecycle Analysis (LCA) for Environmental Assessment

Organizations have need to assess their process for environmental impact for optimizing natural resources inputs, avoiding waste and reducing energy inputs in all stages.

For this LCA facilitates to reduce the environmental loads by assessing at each stage of the product. This technique is for identifying and evaluating the envrironmental impact of producing, using and disposing of a product and can be applied to building, operating, docommissioning a plant or facility. The main objective is to focus (a) Primary inputs and wastes associated with reuse, recycling and ultimate disposal. LCA is an instrument that follows an internationally accepted methodology, to support decisions to improve environmental performance in terms of reducing the environmental load :

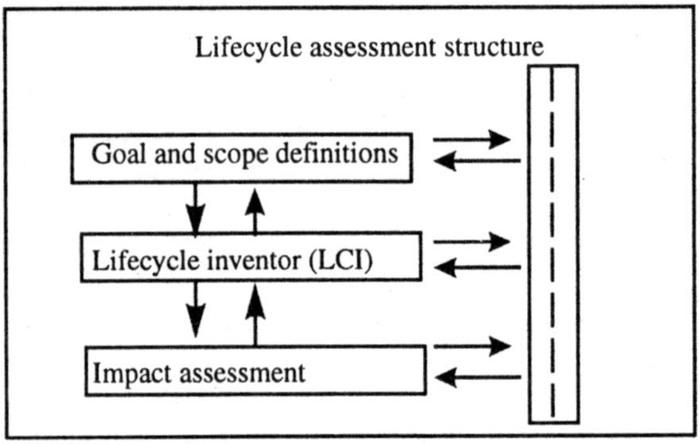

Goal and Scope Definition

- Compare processes in a company that generate products having the same function, in order to determine which is better from and energetic and environmental stand point.
- In addition, it refers the *functional* unit in which flows of all raw materials, resources and emissions in a production system are referred to better comprehend the information provided and facilitate comparisons between system that perform the same function, then looks in to system boundaries
- *System boundaries* begins with all the raw materials taken from the environment and ends with the outputs released back to the environment :

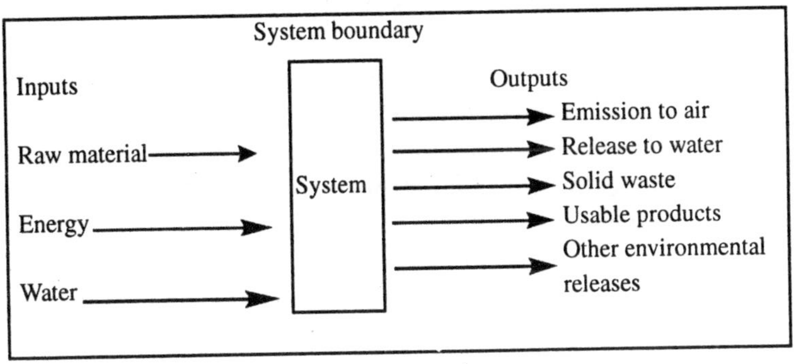

Here LCA begins with the extraction of the raw material and energy from the environment and ends with the return to the environment of all materials used. Now the analysis is to identify the direct environmental loads related to the main production process and indirect loads related to the production of auxiliary materials, energy, plants and the machinery used in the main process to obtain co-products or to obtain production processes that involve recycling materials in order to reduce the emissions and residues to the environment.

- *Critical review* has been carried out till the results are clear and coherent in accordance to International standards.

Lifecycle Inventory

It is an operation of gathering and organizing the flow values exchanged among the individual operations in the production chain.

Impact Assessment

It is to attribute the data regarding emissions and consumption, identified through the inventory, to specific categories of impact assessment can be done by—

- *Defining the categories*: All environmental categories to be considered in the assessment are selected to meet general requirements such as completeness, independence, practically and primary requirements such as human health, environmental protection and depletion natural resources.

- *Through Classification:* The inventory data are distributed among the various categories identified, so that each substance are released into the environment is assigned to categories of environmental impact.
- Each classified substance is *characterized* by attributing an equivalency factor. This makes it possible to add the contribution of each substance to the others, in order to determine the overall effect of the process examined for each category.
- *Through valuation,* the rusults obtained during the characterization stage is linked to corresponding, judgements as to the relative seriousness of impact for each category.

Interpretation and Improvements

Interpretation aimed at identifying, qualifying, checking and evaluating the conclusions of the inventory and impact assessment stages. Improvement completes the cycle of analysis and assessment of environmental impact and makes it possible to direct production towards a stated objective of eco-sustainability, as well as eco-efficiency.

Conclusion

Creating EMS is essential for all the organization in order to improve the control of environmental impact. Due to methodology of LCA and organization can induce confidence in the end-user that the product and services are manufactured with the objective of reducing the negative impact on the environment. It may lead to gain media support and enhance brand name.

Now-a-days EMS is comparatively a new subject and the

information generated about various environmental issues is still quite meagre. The perceptions of EMS are still fractured and incomplete. More research is needed to understand the nature of the problems involved. There is a need to document successful institutional innovations and disseminate this information for the benefit of others.

REFERENCES

A.K. Saxena, "Hazardous Wastes: Identification, packaging, labeling and tracking, productivity, Vol. 33, pp. 638-643, January- March 1993.

Anna Morgante, Andrea Raggi and Luigiapetli, "Instruments for the asssessment of business environmental performance", the current state of business discipline, Vol.4, pp. 1809-1836, 2000.

Dr. M.N. Madhyastha, *Prospects & Problems of Environment—Across the Millennium*, Daya Publishing House, Delhi, 2003.

Prabha Chandra, *International Eco Tourism—Environmental Rules & Regulations*, Kanishkha Publishers & Distributors, New Delhi, 1st Ed. 2003.

Ramaraju Thirumalai, *Resource Management for Sustainable Development*, Himalaya Publishing House, First, edition 2003.

Subhabrata Bobby Bannerjee, "Corporate Environmental Strategy: A review the current state of business discipline, Vol. 4, pp. 1809-1836, 1st Ed. 2000.

S. Maudgal & M. Kakkar, "Environment-Technology Linkages", Productivity, Vol. 33, pp. 630-638, January-March 1993.

Thomas H. Davenport, *Information Ecology*, Oxford University Press, Inc., 1997.

V.K. Prabhakar. *Global Environmental Issues*, Anmol Publications Pvt. Ltd., 1st Ed. 2001.

Vivekananda Kendra Prakashan, Akshaya Vikas—Sustainable Development, Vivekananda Kendra Prakashan Trust, Chennai.

20

ECO-LABELLING AND GREEN MARKETING IN INDIA

Emerging Issues

Dhulasi Birundha Varadarajan and *M. Rajadurai*

Introduction

The concern of the society for the natural environment and drive for making a green earth business has led to emergence green activities that include green marketing, green advertising, green products, green purchasing, green supply chain and green shopping. Some businesses have been quick to accept concept like environmental management systems and waste minimization and have integrated environmental issues into all organizational activities. Some evidence of this is the development of journals such as business strategy and environment and green management international, which are specifically desired to disseminate reseach relating to environmental behaviour of business houses. Thus terms like green market and environmental marketing appear frequently in the popular press.

Eco-labelling

An Eco-label is a label which identifies overall environmental preferences of a product or service with in a particular product/ service category on lifecycle consideration. An Eco-label is awarded by an important third party to product or services, which are found to meet, establish environmental leadership criteria. In India, we come across many advertisments which make false claims about their product and we term this as self-styled green claims which have no authencity.

The objectives of eco-labelling according to ISO are that through the communication of verifiable and accurate information, that is not misleading, on environment aspect of products and service to encourage the demand for the and supply of those products and services that cause less stress on the environment, there by stimulating potential for market driven continuous improvement.

ISO has identified the three major environmental labeling :

- Type I—environmental labelling (i.e. eco-labels).
- Type II—self-declaration clients; and
- Type III—environmental declaration like report cards/ information labels.

In a typical Eco-labelling programme product categories and criteria as established by an independent organization and technical advisory group. Generally, once a category is chosen, lifecycle review of products is conducted and for products this review includes the process of raw material extraction, manufacturing, process, distribution process and criteria, use and final disposal of product. The basic differenting parameter like energy use and toxicity are given maximum importance as they are directly linked to the environment organizations, which want to participate in such eco-labelling programmes are required to pay the licensing for the seeking permission use the programmes' distinctive eco-label symbol for a specified period. It must be noted that use of equal lable is restricted to the approved products and is usually monitored by managing agencies.

The basic objectives of eco-labelling are :

(a) Protecting the environment.
(b) Encouraging environmentally sound innovation in industry.

The guiding principles of eco-labelling are:

1. Voluntary participation.

2. Compliance to environmental and other legislations.
3. Consideration of fitness for purpose and level of overall perfomance.
4. It must be based on sound scientific and engineering principles.
5. Criteria must distinguish leadership.
6. Criteria must be credible relevant attainable and measurable.
7. The programme should be operated independent of vested commercial interests.
8. The programmes must be legitimate.
9. It should be open and accountable.
10. The content of flexibility must be there in all eco-labelling programmes.
11. It must be consistent with ISO-14024 and ISO-14020 guidelines.

The eco-labelling programme India is eco mark India. It provides manufacturers and importers with on incentive to reduce the adverse environmental impacts of their products; reward genuine initiatives by companies to protect environment; assist consumers in becoming environmental responsible in their daily lives by providing them with information regarding environmental impacts that a product can have, thereby improving their purchase style, encourage citizens to purchase products that have fewer environmental impacts and ultimately improve the quality of the environment and encourage sustainable management resources.

Green Marketing

Green marketing incorporates a broad range of activities including product modification, changes to the production process, packaging changes, as well as modifying advertising. Yet defining green marketing is not a simple task. Indeed the terminology used in this area has varied, includes green marketing, environmental marketing and ecological marketing.

Green or environmental marketing consists of all activities designed to generate and facilitate any exchange intended to satisfy human needs of wants, such that the satisfaction of these needs and wants occurs with minimal detrimental impact on the natural environment. In fact, green marketing is the kind of marketing focused on the condition of the earth and human life too (Polansky 1995). So it teaches us to manage the marketing process for satisfying the needs, wants and demands of customers and society in both profitable and environmentally sustainable ways.

Conventional marketing involves selling products that satisfy consumer needs at affordable prices. Green marketing must also satisfy consumer needs at affordable prices, but green marketing has the additional challenge defining what is a green product and developing and selling product that consumer will like. Green products balance environmental and compatibility with performance, affordability and convenience. They are typically durable, non-toxic, recyclable and are often made from recyclable materials. Green products have minimal packaging; little embodied energy and should carry low environmental impact.

The Government of India and few other organizations have made efforts to promote green marketing. In 1991, the Government of India instituted a voluntary scheme to label customer products as environment friendly and to label a green product, the Government adopted a sign of Eco-mark. The American Marketing. Association (AMA) held the first workshop on ecological merketing in 1975. The proceedings of this workshop resulted in one of the first books on green marketing entitled "ecological marketing (Henion and Kinnear 1976) the work on green marketing was published in 1993. The norms for eco-labeling are established by the Eco-mark. Technical committee of the Central Pollution Control Board. Ministry of Environment and Forest, Government of India with Bureau of Indian Standards (BIS) as the implementing authority for the Eco-mark. Eco-mark was to be used in conjunction with ISI mark.

There are some visible instances in which Indian organization have made efforts to promote green marketing (Das & Vikasnath 2004).

1. After achieving the distinction of becoming the first Indian passenger car company to be awarded ISO-14001 certification in December 1999 for its environmental management systems. Maruthi has taken another step towards its commitment for a cleaner and greener society by launching an innovative programme called greening the supply side management (GSCM). Through this programne, Maruthi is going to help its suppliers ensure environment friendly methods of their operations, products and services.
2. The Government of India adopted Euro norms for available fuel quality and methods of testing Euro-I norms in India are known as BHARATH-I norms and Euro-II norms are known as BHARATH-II norms. The central motor vehicle rules 115(11) 1989 says that pollution standards laid down for carbon monoxide and inspected in petrol vehilcle should be as under:

    - 2 and 3 wheelers—idling carbon monoxide emission by volume should be 4.5 percentages.
    - 4 wheelers—idling carbon monoxide emission by volume should be 3 percentage. Various norms have been laid down for other taxic gases like sulphur dioxide, nitrous oxide, carbon dioxide etc.

3. There are 16 products categories currently listed for consideration for eco-marks, like soaps and detergents, paper, food items, lubricating oils packaging materials, aerosol propellants and many more.
4. The first eco-mark was awarded to Godrej product, EZEE a liquid detergent for working on special delicate fabric.
5. Eco mark was awarded to Bharath Paper Limited, Bilaspur for the two types of writing and printing paper.
6. Apart from the some hotels in India are opting for "eco tel" a third party certification programme for eco-friendly hotels.

Eco-marketing in India under WTO Regime

One of the items on which negotiation was agreed upon in Doha ministerial, conference in 2001 was reduction or appropriate elimination of tariffs on environmental goods and services. As of now, there is no internationally accepted definition as to what could be classified as an environmental goods. Nonetheless generally the terms environmental goods and services (EGS) are used for pollution control, reduction, clean up and waste handling services and a growing range of other environmental services (Sandeep Singh, 2003).

Private sector estimates indicate that the global market is US $ 2.85 billion and it is gorwing at a rate of 7 per cent per annum. Recording this the Government of India has already permitted 100 per cent Foreign Direct Investment in this sector.

The global environmental market is clearly dominated by companies from the industrialized world. The offers of trade liberalization that have surfaced in the ongoing WTO negotiation so far, are not necessarily those, which would benefit India. Negotiations at the WTO are taking place in two different bodies. While the issue of goods is being negotiated in Negotiation group on Market Access for Non-Agricultural Product (NGMA) the issue of services being negotiated in the framework of the Council for Trade in Services (CTS).

Lists of environmental goods that have been prepared by the organization for economic cooperation (OECD) and the Asia Pacific Economic Co-operation (APEC) secretariats. The OECD list is broader in scope and includes cleaner technologies and products as well as chemicals apart from pollution control equipment. The APEC list is more practical from trade negotiation point of view. It includes only equipment that can be clearly recognized by the Harmonized System (HS) code for the purpose of the international trade India is net importer of products mentioned in the two lists APEC and OECD.

Trade in Environmantal Services in India

The w/120 list of GATS classified environmental services as:

- Sewage services,
- Refuse disposal services,
- Sanitation and similar services,
- Other services like cleaning services of exhaust gases, noise abatement services, nature and landscape protection services and other environmental services.

WTO members can base their scheduled commitments under these four environmental sub-sectors. Several WTO members have argued for up dealing w/120 classification to include services related to recycling and protection of air quality. The European commission (EC) has also suggested inclusion of "water for human use" and wastewater. It is important to note here that as per GATS Article 1(3), the general obligation including Most Favoured Nation (MFN) treatment do not apply to services supplied under government authority, neither on a commercial basis for in competition with other suppliers.

Obstacles in Green Marketing

Targeting and reaching consumers in a developing country like ours is an extremely difficult task even today. It is because the population is still largely rural, poor and has less to various media than in more developed countries. As a result they have often prefer low quality and sometimes hazardous products to full their needs rather than take care of their environment. On the other hand how industries develop, advertise, manage and promote environmentally friendly products is related to both internal firm characteristics and external factors (Polansky, 1995). Internally the firm needs technical expertise and a supporting culture to support green product development and externally, it requires development and externally, it requires markets where it can sell its product.

Conclusion

The green marketing was developed but it never picked up. The main reasons may be that people are not aware of this concept or they have been made to understand the importance of this concept.

A product with the Eco-market is a green product. However, how often a green product is seen in the market or how often a customer asks for a green product like what is the response to green concerns and how Indians perceive it are the major concern today before the academician, industry, consumers and other stakeholders.

Now there are many companies and institutions that have started thinking about their environment and frequently expressing their concern scope of green marketing by all organizations irrespective of their size, ownership and the types of products they manufacture. This will not only be the part and parcel of our effort to make India as a developed and pollution free country but also enable our business houses to be truly global to complete in the ongoing global competition.

The negotiations in environmental goods in the WTO are not linked to those environmental services. It must be recognized that merely opening up the markets will not solve problems. Ensuring flow of investments and funds is important in achieving these objectives as also the inflow of technology. Thus there is need for appropriate negotiating strategies to ensure a judicious balance between the economic and environmental interests.

REFERENCES

Das, R.P. Vikasnath (2004) *Green Marketing in India - An Emerging Issue.* http:/ / vikasnath. Com/article2 .htm.

Henion, Karl E. and Thomas C. Kinnear, (1976)"A Guide to Ecological marketing. in Karl E Henion and Thomas C. Kinnear (Eds). *Ecological Marketing*, Columbus. Ohio: American Marketing Association.

Polonsky, M J (1995): In Corporating the Natural Environment in Corporate Strategy: *A Stakeholder Approach Journal of Business Strategies*, Vol.12, No.2, pp. 151-168.

Sandeep Singh (2003) "Green Business Prospects, Environmental Goods and Services at WTO" *EPW*, Vol 38, No. 31 August 2, 2003, p. 3240-42.

21

ENVIRONMENTAL ISSUES IN PRODUCT DEVELOPMENT

A Special focus on social accountability 8000 (SA8000)

D. Ayub Khan Dawood

In the 1960s industry reluctantly acknowledge nounting ecological problems. In the 1970s regulations (mostly in the USA) forced manufacturers to abandon paths of resistance and reassess their records. In the 1980s scientific revelations about ozone depletion and global warming as well as tragic acccidents, brought a worldwide shift in corporate attitudes from reassessment to soul-searching.

In the 1990s industry's desire to protect the environment for future generations has been coupled with an even more powerful force for environmentally correct practice: the bottom line. The following are reasons why environmental issues (both threats and opportunities) have become a strategic priority.

Why Environmental Issues should be a Strategic Priority

1. The market opportunities are enormous. Industry may have created many of the problems the world faces today, but it also holds the key to many solutions. Gustave Speth, former head of the World Resources Institute, speaks of the need for "environmentally benign technologies as wholesale changes being in agriculture, manufacturing, transport and construction.
2. Stakeholders want the facts. The public's appetite for information on corporate Environmental, Health and Safety

(EHS) performance continues to accelerate. And it is not just activists who want to know; the list includes customers, bankers, shareholders and community groups. Ciba-Geigy experienced this phenomenon first-hand in the USA when community action, prompted by a series of problems associated with the plant in Toms River (New Jersey) prevented it from expanding operations there.

3. EHS costs are skyrocketing ... for some time; the chemical industry has battled the rising costs of EHS commitments. According to a Booz, Allen & Hamilton survey, however, companies in many other industries are now reporting spending an average of 2 per cent of sales which they consider extremely burdensome.
4. Customers won't foot the bill. Many business report they are absorbing most of the additional costs of environments protection. In many case, they have no choice because customers are unwilling to help pay the bill.
5. Polluters must accept strict liability. Regulations worldwide are assigning strict liability to the polluter. In the USA, superfund imposes strict liability on companies for the clean up to hazardous wastes on their sites, whether or not they created them. The European Community has developed a policy on corporate liability based on the "polluter pays" principle, which imposes strict liability based on waste producers for any damage caused by their waste during production, transport or unauthorized disposal.
6. Liabilities for criminal and civil damages are a growing threat. When Arthur D. Little polled the American public in 1991, it found that 84 per cent of the respondents believe damage to the environment is a serious crime, whereas 40 per cent rated insider trading as a major offence. Three out of four respondents also felt corporate executives should be held personally liable for damage their firms cause to the environment.
7. Pollution prevention can pay. Companies with successful

corporate waste-reduction programmes, such as Dow Chemical and 3M, stress the economic pay off of reducing waste and pollution. Green improvements tie nicely in to lean factory strategies. For example, using raw materials more efficiently and recovering valuable elements from wastes will result in less overall pollution to treat and discharge and in possible savings.

Green Design Considerations for Product Development

A few environment-friendly design considerations being practiced in industry can be summarized as below:

1. *Minimizing the Materials Quantity:* If any single factor can be isolated as being the most important in meeting environmental objectives, it is the reduction of energy consumption. For example, a full lifecycle analysis of a vehicle may show that using an energy intensive material such as aluminium can so reduce the weight of the vehicle that savings in the energy used during the vehicle's life far outweigh the additional energy used in manufacturing the material.
2. *Minimizing the Materials Variety:* It obviously simplifies ordering and stock keeping, often with some saving in waste materials. It also simplifies recycling, as there are fewer materials to separate and sort.
3. *Assembly Simplification:* A design that is easy to assemble generally saves money for the manufacturer. Also it is environmentally beneficial as it is likely to be simple to disassemble at the end of the product's life. A design that is to assemble and allows good access to components also facilitates servicing and repair thus helping to extend product life.
4. *Product Upgradability:*The production rate of computer monitors has been increasing by 15 per cent a year and is expected to continue to increase over the next ten years.

The total number of computer monitors being scrapped per year worldwide is predicted to rise to 30 million units by the year 2005. This is because the technology in this field is improving at a rapid pace. But as a matter of fact, with products like these, the rate of technological development and the increasing demands of users does not require the replacement of every component and thus protects Environment.

5. *Recycling and Re-use:* Recycling seeks to recover useful raw materials. There are three basic design objectives in facilitating recycling: (i) to ensure that different materials can easily be separated; (ii) to enable the easy identification of different materials; and (iii) to avoid as far as possible difficulties from the cross-contamination of incompatible materials.

Social Accountability 8000 (SA 8000)

SA 8000 is an international standard for assessing social accountability in the workplace. It is a comprehensive, global, verifiable standard for auditing and certifying compliance with corporate responsibilty that the workplace be managed in such a manner that basic human rights are supported and that the management is prepared to accept accountability. Social accountability is a broad term created as business response to a host of operational factors affecting a company's diverse stakeholders including employees, the community, governmental and non-governmental organizations, management, owners and stock holders. Social accountability expectations can vary from one company to another, fair compensation, reasonable working hours, a safe and healthy environment and prohibition of child labour and forced labour are common to all codes. Some codes add requirements for non-discrimination freedom of association, disciplinary practices, or management system controls, etc.

SA 8000 is applicable to both small and large companies that want to domonstrate to customers and other stakeholders. SAI regulates the certification of companies that have expressed the desire to become SA 8000-compliant. Independent auditors, accredited by SAI, carry out certification. Social accountability International (SAI) developed SA 8000 standard in 1997. SAI is a non-profit organization dedicated to the development, implementation and oversight of voluntary verifiable social accountability standards.

SA 8000 is based on 12 International Labour Organization (ILO) conventions, the United Nation's Universal Declaration of Human Rights and the UN Convention on the Rights of the Child. These are:

ILO Convention 29 and 105 (Forced and Bonded Labour)
ILO Convention 87 (Freedom of Association)
ILO Convention 98 (Right to Collective Bargaining)
ILO Convention 100 and 111 (Equal Remuneration for Male and Female workers for Work of Equal Value; Discrimination)
ILO Convention 135 (Worker's Representatives Convention)
ILO Convention 138 and Recommendation 146 (Minimum Age and Recommendation)
ILO Convention 155 and Recommendation 164 (Occupational Safety and Health)
ILO Convention 159 (Vocational Rehabilitation and Employment/Disabled Persons)
ILO Convention 177 (Home Work)
Universal Declaration of Human Rights UN Convention of the Rights of the Child.
United Nations Convention to Eliminate All Forms of Discrimination Against Women.

The standard also requires compliance with prevailing laws, regulations and other applicable requirements such as contractual agreements. Some of the Indian Government Acts parallel to SA 8000 are:

Abolition of Child Labour Act
Payment of Minimum Wages Act
Factories Act
Industrial Dispute Act Wages and Equal Remuneration Act

SA 8000 and ISO 9000 and 14001

Though it is modelled on the well established ISO-9000 and ISO-14001 standard for quality and environment management systems, SA 8000 prescribes specific performance standards. While many clauses of SA-8000 such as preventive and corrective action, management review, planning, implementation and measurement of effectiveness, control of suppliers, records and the need for verifiable objective evidence are similar to the ISO-9000 quality standards.

SA-8000 is a consensus standard and specifies minimum requirements and audit methodology to evaluate workplace conditions including child labour, forced labour, health and safety, freedom of association, discrimination, disciplinary practices, working hours, compensation and management's responsibility to maintain and improve working conditions.

Benefits of SA 8000

Benefits for Workers, Trade Unions and NGOs

- Provides better work conditions and work environment thereby improves the work life of employees,
- Provides opportunities to form trade unions and bargain collectively,
- It educates workers about core labour rights,
- Safe work conditions leads to fewer accidents,
- Improves communication with management and hence reduced grievances,
- Provides opportunity to work directly on labour rights issues.

- Improved work practices leading to increased productivity, higher economic growth and new job opportunities.

Benefits for Organizations

- A credible and effective way to put company values and social responsibility into action.
- Organisation's brand and reputation enhanced.
- Improved employee morale results in talent recruitment, retention and performance.
- Improves productivity and quality and results in lesser workdays lost due to accidents and reduced insurance bill.
- Better supply chain management and performance.
- Harmonious management labour relations and therefore, better industrial relations.

Benefits for Consumers and Investors

- Consumers get useful information about the socially responsible organizations.
- It helps in identification of products, which are made under humane conditions.
- Consumers get trustworthy information for making ethical and principled purchasing dicisions.
- Identification of companies committed to ethical sourcing.
- Broad coverage of product categories and production geography.

SAI acts as the accreditation body and licences qualified auditors to certify workplace compliance and monitor conformity to the standards. SAI regularly audits the auditors. It does not audit factories or farms directly. It imparts training to auditors, workers, managers and suppliers for effective implementation of the system. Accreditation is valid for a span of three years with surveillance and observation audits every six months, covering each/any country where the registrar audits ten or more facilities. SAI certifies individual auditors. At present, the certification facilities are available in more than 30 countries or five continents and across 22 industries. Some of the accredited auditors are:

SGS International Certification Services Ltd.
Det Norske Veritas (DNV)
Bureau Vertias Quality International (Holding) S.A. (BVQ1)
Underwriters' Laboratories (UL)
Intertek Services (ITS)

Johnson and Johnson's Neutral Environmental-impact Policy

Johnson & Johnson has written a neutral environmental-impact policy that is accompanied by specific packaging guidelines. Both these documents establish procedures for estimating the environmental effects of new products, packaging and manufacturing processes before they are in use. The procedures are as follows:

- A comprehensive survey of all possible materials for products and packaging and the production steps is conducted to determine the environmental releases associated with materials selection or manufacturing processes.
- Managements are responsible for communicating any issues that may be identified from these assessments. They must also propose ways to deal with the problems.
- Technical staff members are required to prepare detailed studies and recommendations to managements that outline the technological options to eliminate, reduce, substitute or recycle undesirable substances.
- Packaging guidelines, in order to priority, include the following:

  * No packaging;
  * Minimal packaging;
  * Consumable packaging;

* Returnable or refillable reusable containers;
* Recyclable packaging;
* Recycled content.

- Environment-impact assessments are also part of launch procedures for new products as well as of regular assessments of existing ones.

Conclusion

Presently, we are passing through a kind of environmentally driven revolutionary age. Around the world, enormous resources are being devoted in laboratories and industries to seeking ways of minimizing the environmental damage. As a result, a widerange of new technologies are being generated. This is driving the designers and manufacturer to develop new products, processes and systems, which are not only environmentally friendly but also less energy intensive and cost effective. It is hoped that the future industry will be more energy-efficient, more economically advantageous, less polluting and thus inherently sustainable.

REFERENCES

Burall, Paul (1996): Product Development and the Environment, The Design Council, Gower, Hampshire, UK.

Clegg, A.J., D.J. Williams, and R.U. Soy (1995): Production Planning for Companies with Remanufacturing Capability, IEEE Int. Symposium on Electronics and the Environmental, pp. 186-191.

Gurjar, B.R. (1996): Increasing Global Temperature: Its possible consequences. *Journal of Indian Association of Environmental Management.* Vol.23, 36-37.

Mackenzie, Dorothy (1991) *Green Design: Design for Environment,* London, and Laurence King.

Misra, K.B. (1996) *Clean Production, Environmental and Economic Perspectives*, Springer-Verlag, Berlin Heidelberg, New York.

22

GREEN MARKETING

Environment and Economic perspectives

D. Ayub Khan Dawood

Green Marketing has been synonymously used with terms like environmental marketing and ecological marketing. Polonsky defined green marketing as "all activities designed to generate and facilitate any exchange intended to satisfy human needs or wants, such that satisfaction of these needs and wants occurs with minimal detrimental impact on the natural environment." Although there is no one universally accepted definition of green marketing, the term is used to describe any merketing activity of a firm that is intended to create a positive impact or to reduce the negative impact of a product on the physical environment and as a result capitalize on consumers concerns about environmental issues. It ecomapasses everything from using recycled material in making a product to changes in production process, product modification, packaging changes to claims in advertising.

With the environmental movement gaining strength, consumes have started showing an increased willingness to pay a premium to support truly environmentally friendly products and companies. The findings of Green Gauge Report a research conducted on American consumers showed that about 10 per cent of the U.S. adult population was willing to pay 7 per cent more for ecologically friendly products, while 5per cent of the population who were concerned and reasonably well off were willing to pay as much as 20 per cent extra for green products. In a study conducted in 16 countries in 1992, more than 50 per cent of the consumers in each country except Singapore indicated that they were concerned about the environment. With such dramatic changes in consumer behaviour firms have had no option other than to think and practice green. By offering green products, companies can enhance their profits, market share, contribute towards their corporate image building and generate a lot of goodwill. Organizations are also going green in their production and marketing efforts as they feel they have a moral obligation to be more socially responsible. As a result of this environmental issues have been integrated in to the firms' corporate culture. Walt Disney World for example has an extensive waste management programme.

Government Role

Government regulations to protect consumers and society have invariably contributed towards the growth and success of green marketing. Government through its various rules and regulations has tried to —

- Reduce production of harmful goods and by-products,
- Control the amount of hazardous waste produced by firms,
- Protect consumers from false and misleading claims,
- Ensure that consumers are able to make more informed decisions, and
- Induce final consumers to become more responsible.

Competitive pressure too has contributed significantly towards the growth of green marketing. Besides these, cost and profit issues have also forced the firms to think green. The re-examination of their production process to minimize their wastes not only helps them in reducing waste but also at times reduce the need for some raw materials. This serves as double cost saving as both waste and raw materials are reduced. With the green revolution gaining strength many organizations have turned over a green leaf. True green work requires companies to practice the tree R's of waste management; *Reducing, Reusing and Recycling* waste.

In India too, companies are changing their image from environment polluting to environment-friendly companies.

- Badarpur Thermal Power Station of NTPC in Delhi is devising ways of utilizing coal-ash that has been a major source of air and water pollution.
- Barauni Refinery of Indian Oil Corporation is taking steps for restricting air and water pollutants.

More Green Product Innovations

Early green products, such as CFS-free aerosols, phosphate-free detergents and cadmium-free batteries, often focused on eliminating a single harmful raw material in the product. Today new green product designs or reformulations of old products are also stressing energy efficiency, recycling and waste reduction.

This list highlights recent environmental innovations in products and services:

- Photocopiers. In 1993 Ricoh introduced a prototype reverse-copier machine. When a piece of paper. Ricoh management designed the machine as a less expensive alternative to produce recycled paper. It plans to introduce a more sophisticated reverse copier within the next few years.
- Energy-efficient computers. The US Energy Star Computer programme has inspired leading manufacturers to develop computers that will go in to a low-power state, or "Sleep mode", when they are not in use. Acer America, a subsidiary of the Taiwan electronics firm, was one of the first to market with these new machines.

According to the EPA, the Acer systems will save enough electricity to provide power to Vermont, New Hampshire and Maine each year. And they will save personal-computer owners up to $2 billion in annual electricity bills.

- IBM has its green PC, the PS/2E, which will reduce energy consumption over 75per cent compared with the regular PC. Other green features include designs for disassembly and recyclability and a cover containing recycled materials. Intel's new chips will also be designed to go into the sleep mode when not in use.

- Bioplastics. Many companies in the USA and Europe are investing large amounts of money in research and development of polymers derived from plant matter rather than from petroleum.
- Novon Products Group, a division of Warner-Lambert, invested more than $100m over the past decade to develop a 100 per cent corn starch-based polymer resin. Cargill is building an $8m plant in Minnesota to produce a lactic acid-based polymer resin employing a sugar-fermentation process. ICI has also introduced a fermentation-grade plastic called Biopol.
- Eci-fibre. Wellman has introduced a velour fleece fabric made from 100 per cent recycled post-consumer plastic beverage containers. It is the first commercial textile fabric made from this waste source. The product was developed in partnership with customers Dyersberg Fabrics (Dyersberg, Tennessee) and outdoor-sportswear manufacturer Patagonia.
- Jeans to paper. Denim scrap from Levi Strauss is being turned into a unique paper product. Working with Watson Paper of Albuquerque, New Mexico, Levi Strauss has supplied the paper maker with a steady supply of scrap material that is turned into a paper that resembles faded blue jeans. The company has used the paper in a variety of way, including business cards for employees and corrugated boxes.
- Reusable envelops. Several North American Banks send their bills to credit card customers in two-way envelopes. The flap on which the bank's address is printed can be inverted and resealed. The envelope is made with recycled paper, water-soluble glue and a window of glasses (a wood pulp product).
- Recycled rubber boots. Dunlop, Britain's largest

manufacturer of Wellington style rubber boots (known as Wellies), has built a plant to recycle worn-out wellies and is setting up collection blanks. These boots, made of PVC, can be stripped of cloth, chopped and remade more than once.

Four Steps to Encourage Suppliers to be Green

Companies can approach the greening of their supply chain in a variety of ways. Because this is such new territory, few firms have developed a formal strategy.

The experience of companies thus far suggests a four-step framework for influencing and improving a supplier's performance. Examples of firms in this checklist are drawn from "The E-factor" by Joel Makower, and a new report, buying into the environment: Guidelines for integrating environment into purchasing and supply", co-authored by Business In the Environment (BIE), the British Chartered Institute of Purchasing and Supply and KPMG Peat Marwick.

1. *Establish and Communicate Clear Standards :* Campbell's Soup insists that all the glass jars it receives from suppliers have a post-consumer recycled content of 25-27 per cent. Suppliers are also required to submit a corporate policy statement that specifies exactly what proportion of recycled glass is used by their business.
2. *Develop Profiles of Suppliers :* Questionnaires and surveys are favoured to find out more about what suppliers are doing or will do to meet a customer's environmental standards. Questionnaires should be very specific, requiring yes or no answers. The two purchsing groups at Sun Microsystems—components and office supplies—worked with the environmental, health and safety department to develop a supplier survey that would identify where ozone-depleting solvents were being used in the manufacture of components.
3. *Validate Performance :* In some instances, audits of

performance may be appropriate. Those that include visits to a supplier's site may be conducted as part of a rating exercise, to judge compliance with customer standards and to suggest areas for improvement.

Smith Kline Beecham uses an in-house team to audit its waste contractors with site visits. Ace Hardware's paint division sent a letter to vendors with a 16-page questionnaire along with its environmental policy statement.

4. *Provide Support* : Companies that do establish strict standards are most likely to see their suppliers meet them halfway if they reach out with support. For example, after Northern Telecom completed a worldwide survey to determine supplier progress in eliminating the use of ozons depleting solvents, it began to hold outreach seminars and workshops.

Conclusion

Green Marketing covers more than firms marketing claims. In fact environmentalists and regulators became concerned when companies went overboard in their use of terms like recyclable, degradable and environmentally responsible. However, with the beginning of a new century environmentalism appears to be moving into a more mature phase. Green concerns today are broader, deeper and more sophisticated.

However, while firms must bear much of the responsibility for environmental degradation it must be remembered that it is the uncaring customer who chooses to dispose of his waste in an inappropriate manner. While firms can have a great impact on the natural environment, the responsibility should not be theirs alone. Ultimately, marketing requires that consumers demand a cleaner environment and show willingness to pay for it, possibly through higher priced goods, modified lifestyles or even governmental intervention. Until this happens it will be an onerous task for firms alone to lead the green marketing revolution.

REFERENCES

Dhingra, Atul and G. Arya (2000). Green Marketing - Advantages and Challenges, jims 8m, 5(3) July - September. N.L.Gurjar, B.R. (1996) : "Increasing Global Temperature : Its Possible Consequences". *Journal of Indian Association of Environmental Management,* Vol. 36-37.

Mintel (1994): *The Green Consumer*, London: Mintel, London.

Kotler, Philip and G. Armstrong (1999). *Principles of Marketing.* New Delhi: Prentice Hall.

23

NEW DIMENSIONS OF TOURISM COOPERATIVE ORGANIZATIONS IN PROMOTING ECO-TOURISM IN DEVELOPING COUNTRIES

Samwel Kakuko Lopoyetum and *P. Selvamani*

Introduction : An Overview

In the era of economic liberalization and globalization, almost all the activities touching on the tourism are extremely very important and highly essential to be streamlined in order to protect the environment and provide adequate benefits of the local community. Tourism has socio-natural and socio-cultural effects at local, Regional, State and National levels. Moreover, tourism is an important aspect of human life. As an economic activity, it is a part of the services sector of an economy and is a big source of employment generation and foreign exchange earning in many developing countries. Tourism could be for sight seeing, holidaying, education, business, medical treatment, rejuvenation of mind and body, etc. In many countries tourism is treated as an industry. It is indeed difficult to pin-point the period of human history when tourism began. But, one can easily estimate that, tourism could be as old as human race itself. Along with mankind, tourism has also registered a phenomenal growth over the centuries. Now, tourism enjoys an incredible status of being one of the largest and fastest growing sectors of the world. Tourism which cuts across caste, colour, creed, race, language, etc. People from all walks of life, of all ages and of all income groups experience the spirit of tourism at least once in their lifetime. Worldwide, tourism has achieved rapid strides and growth due to its amazing capacity to earn foreign exchange, generate employment opportunities and inculcate sense of love and brotherhood in the mind set of the people.

Relevance of Tourism Cooperative Organizations

Cooperative tourism can promote eco-tourism in all its dimensions and even in the future perspective. They can formulate suitable sustainable environmental management and generate tourism revenues and employment in both the rural and urban areas. The rural communities living in the major spot surrounding tourist areas can protect the natural environment and provide guidance to the operations of tourism business. Cooperative tourism can easily provide adequate income and diversification to rural communities. Most of the income generated from the outside visitors can straightaway stay within the local economy. In addition to this merit, the cooperative members stand to gain from direct sales and transactions with the tourist visitors, many indirect benefits are realised from tourism. Visitors contribute to the tax revenues collected and can influence the quality of life by financing community facilities such as swimming pools, golf courses, restaurants , shopping facilities and other permanent infrastructure facilities. The community events, intended for tourists, also attract local residents. Many potential industries prefer locations that provide high quality of services, prompt services, quality services and recreational resources. If the community is a pleasant place, the visitors may become a permanent resident. Cooperative tourism do provide products and services to tourists who creates new jobs in the community and these specific jobs can satisfy the need and wants of the local community and even the dual-earning families for part-time or seasonal employment.

Cooperative tourism requires local community participation and gain support by informing and educating the cooperative members. They can introduce this idea to the Chamber of Commerce, at civic clubs, at city council, or at other community organization meetings. The community and its leaders can broaden support for cooperative tourism by securing financial commitments from public and private sources. As far as cooperative organizations is concern, it is important to involve all facets of the members in the decision making process. Everyone and everything the visitor has contact with, must project and promotes something about the members and the local community at large.

Table 23.1-23.3, reveals the growth and the progress of the tourism industry in Tamil Nadu and India. It indicates the tourist arrivals in terms of the demestical arrivals and the foreign arrivals. It reveals an extreme deviation in terms of arrivals and the percentage shares of the Tamil Nadu State with the All-India share. On the other side, they are almost at equilibrium level. However, it is essential to promote both the domestic and foreign tourism in India.

Table 23.1 : Status of Tourist Arrivals in Tamil Nadu

(Figures in lakhs)

Year	*Domestic*	*% in Gr. rate*	*Foreign*	*% in Gr. rate*
1993	142.12	--	4.35	–
1994	160.26	11.32	4.98	12.65
1995	172.14	7.00	5.86	15.02
1996	182.03	5.43	6.14	4.60
1997	189.28	3.83	6.36	3.44
1998	204.13	7.30	6.36	0.00
1999	211.37	3.43	7.22	11.91
2000	229.82	8.03	7.86	8.14
2001	238.12	3.49	7.73	-1.68
2002	246.62	3.23	8.05	4.00
2003	270.17	18.16	9.02	10.75
Average	204.17		73.93	

Source : Director of Tourism, Chennai.

Table 23.2: Foreign Tourist Arrivals to India and the Percentage Share of Tamil Nadu

(Figures in lakhs)

Sl. No.	*Particulars*	*2001*	*2002*
1.	All India	25.37	23.70*
2.	Tamil Nadu	3.10	2.97
3.	% Share	12.2	12.5

Source : Director of Tourism, Chennai.

Table 23.3 : Indian Tourism 2002-2003

Foreign Tourist arrivals during 2002 for India	2370121*
World Tourist arrivals during 2000	698 million
Percentage share of India in World Tourist arrivals during 2000	0.38
World Tourism Receipts (Excluding Fare receipts) during 2000	462.2 (Billion US$)
Estimated foreign exchange receipts of India during 2002	14419.82*
Percentage share of India in World Tourism receipts during 2000	0.66
Estimated number of domestic Tourist movement in the country during 2001	234 million

Source : Director of Tourism, Chennai.

Ways for Developing Cooperative Tourism

The following are few ways and means of developing cooperative tourism:

(a) It is essential to carry out feasibility studies and get proper understanding on the potential for cooperative tourism development in the country;
(b) It is imperative to inform and educate the community/ people to create support for the sustainable growth and development of the cooperative tourism;
(c) Cooperative sector can secure investments from public, cooperative sector itself,
(d) It can effectively manage natural, human, and financial resources for the benefits of the local community; and
(e) They can build an appropriate image for the community and public at large.

Services which can be offered by Cooperative Tourisms

They can render the following services viz.,

- Tourists visiting the State can seek pleasant experience on their trips. The cooperative organizations with a view to providing congenial atmosphere for the promotion of

Tourism can concentrate on quality of tourism services and products.

- They even draw an Integrated Special Tourism Schemes for Tourist development centres with reference to supply adequate infrastructural facilities.
- They can provide accessible tourist attractions and integrated system of physical infrastructure which are stated to be eco-friendly.
- They must promote sustainable Socio-economic development of the area through community participation and action for sustainable tourism and eco-tourism.
- Encouraging private participation in all Tourism ventures and business activities.
- They must promote eco-tourism and manageable level of tourism ventures.

The Imperative of Eco-tourism

Meaning of Eco-tourism: Eco-tourism means undertaking all those tourism business activities which are responsible for better travel to natural areas that conserves the environment and improves the well-being of local people.

The General Goals and Aims of Tourism

The following are the general goals and aims of tourism:

- To promote tourism and eco-tourism as way of life and a basic human right which enriches individuals, family, societies and nations.
- To stimulate communication about the benefits of tourism and eco-tourism as the most prospective economic activity for the local communities and encourage local participation.
- To enhance cooperation and coordination between destinations and the tourism industry with the local and regional levels in all the developing countries.

- To link individual tourism entities and infrastructure development to the benefit of the larger community of the local and regional tourism environment.

Various Principles of Eco-Tourism

The following are the principles of the Eco-Tourism viz. :

- To provide financial benefits and empowerment for local people/community.
- To build an appropriate environmental and cultural awareness and respect among the rural people/community.
- To provide positive eco-tourism experiences for both the visitors and tourist from different areas, hosts and organizations, particularly in most of the developing countries.
- To provide direct financial benefits for the conservation of the natural resources and the local population must gain adequate opportunities and benefits from the exploitation of natural resources.
- To raise sensitivity to host countries' political, cultural traditional environmental, and social climate.
- To support the international human rights and labour agreements among different tourism organizations.

The Benefits of Tourism and Eco-Tourism

Tourism is an important factor in the balance of payment and foreign exchange earned for many of the developing countries. Tourism is an important job creator, employing millions of people around the world. Tourism jobs and businesses are usually created in the most underdeveloped regions of a country, helping to equalize economic opportunities throughout developing nation and it can be used for providing an incentive and opportunity for residents to remain in the rural areas. The tourism industry provides governments with hundreds of millions of dollars in tax revenues for each year.

It is the policy of cooperative tourism organizations to consider their cultural and historical heritage as a valuable resource that needs to be managed in such a manner as to promote conservation, public involvement, research, and tourism development. It is, further essential to extend and promote, the policy of cooperatives to cooperate with the local governments, organizations and individuals in order to improve their resource management and promotion for the benefit of their respective members and non-members. Cooperative tourism organizations authority includes those statutes and policies that address to the cultural and histroric sites, resource, and properties, entering into cooperative agreements and transferring funds in support of such agreements. This is highly imperative in the developing countries and the merits and benefits of tourism and eco-tourism are extremely very large in the long run.

Types of Tourism

(a) *Leisure Tourism:* Tourists visiting tourist centers or spots in Tamil Nadu for attainment of their leisure constitute a major segment of the source of tourists. They usually throng to the hill resorts centres, beaches, waterfalls, zoological parks, bird sanctuaries, botanical/horticultural gardens, etc.

(b) *Pilgrim Tourism:* This is another major segment of tourists visiting Tamil Nadu. They visit places of worship importance according to their faith. But, at times their zest and zeal transcends all barriers and they do not hesitate to visit places of worship of other religions.

(c) *Heritage Tourism:* This group is normally made up of people in their fifties and above, who would like to visit historical monuments.

(d) *Adventure Tourism:* This is the most vibrant group made up of young boys (Youth) in their teens and twenties. They like to whet their appetite for adventure through trekking, air/water sports, etc.

(e) *Business Tourism:* Top executives of different trade and business organizations who arrive in Tamil Nadu to attend seminars, conferences, conventions, and general body meetings, elections, etc. They usually combine tourism with their business trips.

(f) *Medical Tourism:* Tamil Nadu is proud of possessing India's best medical facilities (Allopathy, Ayurveda, Siddha, Unani, etc.). This coupled with Tamil Nadu's traditional hospitality attracts patients from other States and even from neighbouring countries and many overseas countries. After full recovery, the patients as well as their attendants will eventually visit places of tourist interest.

(g) *Fairs/Festivals Tourism:* Large inflow of tourists is usually seen during the major festive seasons like Deepavali, Christmas, Ramzan, Pongal, etc. The exhibitions, fairs (like India Tourist & Industrial Fair conducted every year by Tamil Nadu Tourism Development Corporation at Island Grounds, Chennai), etc. also draw significant number of tourists.

(h) *Academic Tourism and Research Oriented Tourism:* Students, researchers from educational institutions and establishment visit tourist spots as part of their curriculum. The tour, though primarily meant for informative/educative purposes, is later converted as a tourist trip. Students are given consession in entry fee to Poompuhar Tourist Complex, Sound and light show at Thirumalai Mannar Palace, Madurai and Kattabomman Memorial Fort, Panchalankurichi, which are under the control of the Tourism Department.

Organizations Promoting Eco-tourism and Tourism Cooperatives

- *International Cooperative and Associated Tourism Organization (TICA)*

The International Cooperative and Associated Tourism Organization (TICA) is a specialized organization under the International Cooperative Alliance (ICA), which is involved in the tourism aiming at developing tourism at the National and

International level. They are setting up tourism cooperatives in member countries.

The International Co-operative and Associated Tourism Organisation (TICA) is one of the wings of the International Co-operative Alliance. TICA became one of the Specialized Organisations involved in tourism which aim to promote the development of tourism and eco-tourism at the national and international level by implementing the following principles:

- To support all measures taken by public authorities at national and international level, to aid investment in tourism and to make vacations accessible to the greatest number of people possible.
- To complete tourist information by the accurate and objective advertising of available services and to work towards improving the quality of services offered to tourists.
- To provide information to travellers visiting a country where there is a co-operative movement, including details of co-operative events scheduled during their stay.
- To support the development of any organisation which facilitates the opportunity of holiday stay and travel, and offers fair prices and good quality service, to its member organisations.
- To develop inter-cooperative relations and also to help set up tourism co-operatives in member countries of the ICA which are still without such organisations.

Need for Cooperative Research Centre for Sustainable Tourism Development in Developing Countries

The Cooperative Research Centre for Sustainable Tourism was established in 1997 to underpin the development of a dynamic and sustainable tourism industry. Its mission is the development and management of intellectual property (IP) to deliver innovation to business, community and government, enhancing the environmental, economic and social sustainability of tourism. CRC Tourism focuses on four key areas of the tourism market:

- Active conservation and environmental management
- Engineering, design and technology
- Appropriate and suitable policy, events and business management
- Information Technology (IT) and informatics

The major purpose and aim of setting up the Cooperative Reserch Centre for Sustainable Tourism Private Limited is to develop and manage the intellectual property (IP) to deliver innovation to business, community and government, to enhance the environmental, economic and social sustainability of tourism and eco-tourism.

Purpose of Cooperative Research Centre (CRC)

There is need for the cooperative tourism for improving the communication process/system and coordination among the several interests involved in cultural resource management and natural resource use efficiency. There is an urgent need to coordinate projects and programmes of mutual interest and to improve the communication process and coordination which can best be accomplished through the establishment of interest among the cooperative members.

Multi-state Cooperative Society to Promote Tourism—Karnataka Experience

Multi-state cooperative society to promote tourism is doing the work of promoting and developing potential of tourism and eco-tourism in Karnataka, Kerala and Goa. The cooperatives of these States have come together to form a Multi-State Cooperative Society. The Bharthi Tourism Development Cooperative Ltd. (BTDCL) with its headquarters at Panjim in Goa is one of the pioneering attempts in the country to establish, develop and promote tourism business upholding the cooperative principles and practices. The main aim of the organisation is to provide excellent quality or service to people and to promote tourism and hospitality services based on pure Indian culture and ethics. In Karnataka, Bharathi Tourism had developed and promoted the adventure tourism in the initial phase. Stating that Karnataka has a wide range of Western Ghats, rivers and beaches to promote this package tourism and eco-tourism. The BTDCL has focused on the school students and youth to promote tourism and eco-tourism in Karnataka. The cooperative venture will operate tours to and from Kerala, Karnataka and Goa, promote farm, spiritual and eco-tourism, set up information centres at major tourist spots and collaborate with natural resorts, hotels, travel agents and tour operators.

Lessons and Experiences from other Counries on Tourism and Eco-tourism

In many developing countries such as Philippines and Malta, they have "Green Tourism" and in Malta they have bus-drivers cooperatives. Even in Nepal, they have Mount Everest Cooperatives which are selling tourism services to tourists. The Tourism Cooperatives have tour operators or running travel agencies. All these Tourism Business activities must be undertaken in a manner to meet the principles and practices of Eco-tourism. It must conserve the environment and provide financial benefits and empowerment to the local people. They even protect the environment and provide some rights to the local community.

Sustainability of Eco-tourism Cooperatives

It is essential to sustain the growth and development of the tourism and eco-tourism. The following ways and means can be adopted.

- The Government should encourage agencies to incorporate exhorbitant growth of tourism. and eco-tourism cooperatives in their planning processes, including at the regional level, and grassroots level.

- The cooperative organizations and Government must ensure that the growing tourism and eco-tourism strategy shall remain as dynamic, responsive and relevant to future industry and community needs in the new century.
- The tourism cooperative organizations should use the new slogan and attractive logo on destination brochures and advertising on eco-tourism system. The cooperatives must make use of the slogan, logo, and artwork (poster) in the dynamic print media and press familiarization trips. Some of the SHG's can prepare some of the Art materials and logos.
- There is imperative need for the promotion of the community tourism in the developing countries.
- The tourism cooperative can publish the flyer in a effective and efficient local languages even the local languages can be used in the documentation process.
- The Government should take adequate measures to printing t-shirts or billboards with the suitable slogan and logo of sustainable eco-tourism and promote the growth and development of tourism cooperative organizations.
- The Government should take adequate measures to the growth and development of sustainable tourism and eco-tourism through triablazing science carried out in its Co-operative Research Centres and suitable educational establishment.
- Cooperative tourism organizations should enter into maximum and optimum utilization of the available natural, human and financial resources. They must manage all these resources in the most effective and efficient manner. They are the greater importance and component of tourism development.
- Cooperative tourism organizations can provide an opportunity to analyzes information about domestic and international travel behaviour and trends that directs and

drives the programmes and tourism services. This information may provide advertising planning and organizing as well as evaluation of the advertising programmes to reflect the actual position of the tourism industry.

- The cooperative tourism organizations should develop the benefits of its member's particularly rural and urban-based areas, where tourism is often assumes an important source of income, tax revenue, and job creation for the community.
- A comprehensive cooperative tourism development outreach programme with much to effort provide training, assistance, and technical expertise through, assessments, and presentations to help their members, business, and different cooperative tourism organizations can be developed. It must recognize, develop, package, and market their tourism destinations and products in the most cost-effective and efficient manner.
- New cooperative tourism developments on the coastal and islands are subject to increasingly tough controls over design, construction, implementation, executives, environmental impact, energy and water use and especially waste disposal, to prevent affluent runoff to the sea. If this is the situations, it is essential to promote the most sustainable Eco-tourism ventures along the coastal regions of Tamil Nadu and the country at large.

Promotion of Sustainable Agri-Tourism and Rural Tourism

Meaning of Agri-Tourism: Agri-Tourism is the economic activity that occurs when people link their travel and tours with agricultural products, services or experiences, agricultural farms, Estates etc., Agri-Tourism has recently emerged as an important opprotunity for many farm families to diversify their agricultural operations and business activities. Such as on-farm diversification is income generation, as well as increased awareness and education of the non-farming public about the importance and scope of farms and their products/commodities. Gandhigram is endowed with evergreen villages. A village such as Ratnagiri can be adopted as

Agri-tourist and Rural Tourism village. Similarly, other villages can be included in the list of the Agri-tourist and Rural Tourism.

Rural Tourism: Across the world, the trends of industrialization and development have had an Urban Centric approach. Along side, the stresses of urban lifestyles have led to growing interest in the rural areas. In most of the developed countries this has resulted in a new style of tourism of visiting village settings to experience and live a relaxed and healthy lifestyle. In India, also there is growing interest in such rural tourism. Government of India has thus come forward with a scheme under which sanction up to Rs. 50 lakhs for promoting rural tourism in one centre. The benefits of tourism must percolate from urban areas to rural areas, which is bring and create adequate employment and income opportunities for village people including skilled and unskilled persons. Cooperative sector can benefit from this scheme set out by the Government of India.

Even now-a-days, people are turning to the family farm as a tourist destination spots/centres for many reasons such as :

(a) Interest in the natural environment which is actually a desire for peace and tranquillity, or recuperation from illness or sickness.
(b) Disillusionment with overcrowded resorts and cities which are risky and hazardous to the lifes of tourist.
(c) Promotion of the rural recreation activities and infrastructural facilities.
(d) It is an inexpensive getaway to the promotion of tourism and eco-tourism in rural areas.
(e) Curiosity about the farming industry and lifestyle of the farming community.

Methodology of Agri-Tourism and Rural Tourism

The following approaches and strategies can be adopted in the promotion of the above tourism viz.,

(a) *Farm Accommodation :* Several forms of tourist accommodation can be provided on the farm. Some farmers can provide adequate and suitable camping sites for tents or trailers, while others can provide self-catering units, whether in a separate cottage or a site within the farm house itself. These self-catering units with sleeping/living quarters, kitchen and bathroom facilities can be treated as a growth area in the farm holiday market and thereafter expanded to include all the tourism infrastrusture facilities.

(b) *Farm Tours:* Day excursions to farms can provide tourists with an opportunity to learn about the economic and practical side of the farming community. They can learn about different types of livestock, farming systems, and crops, view machinery, and equipments and walk through farmyards and fields. Tea and Coffee plantations are the best site for walking through during right day light seasons.

Promotion of Research Ventures in Agri-tourism Spots

The participants of rural tourism will learn the research-based need for agri-tourism as well as the benefits, challenges and opportunities provided by it and also to conduct a group analysis of a case study to assess what needs to be done to make an agri-tourism business successful and how to do it. Also, how to promote the livelihood and the socio-economic conditions of the agricultural farmers. The connection between cultural and nature-based tourism experiences is a logical one, supported by travel motivational research. The types of experiences sought by cultural travellers frequently overlap with the types of experiences chosen by nature-based travellers. Similarly, visitors who are interested in heritage attractions and natural history are usually also interested in hiking, bird-watching, canoeing and bicycling etc. It makes sense for cultural and nature-based tourism to work together to plan exciting itineraries and packages for the travelling public. Agri-Tourism shall remain significant and viable tourism and eco-tourism venture in most of the developing countries.

Extra Role for Agri-tourism and Rural Cooperatives

They can perform extra services such as —

(a) Organize the purchase of bulk supplies (seed, fertilizers, chemicals, etc.) for their members;
(b) Establish manufacturing or processing plants for value adding of member's products/commodities;
(c) Improve marketing of member's produce by providing: market research and development; packing, grading and transport services; and centralized pricing, promotion and sales;
(d) Offer services like: crop credit to members for farm inputs; technical expertise and advice; and harvesting; and
(e) Agricultural tourism, or agri-tourism, is one alternative for improving the incomes and potential economic viability of small and medium farmers and rural communities.

Strategies for Promoting Eco-tourism and Tourism Cooperative Organizations in Developing Countries

- Promote good management practices and provided economic benefits to the local people who are mainly asked to forego their natural resources utilization.
- The governments of developing countries/nations should take measures to secure mutual understanding and co-operation between the national organizations comprising the cooperative organizations for developing eco-tourism and tourism in a sustainable manner.
- The cooperative organizations and concerned authorities can coordinate its members' operations in order to make tourism and its benefits available to as many people even in the remotest of the rural areas.
- The government and cooperative organizations should encourage the regular exchange of date, information, fact and figures about tourism and eco-tourism business activities and their personnel and services among the cooperative tourism organizations in the developing countries.
- The tourism cooperative organizations and government

approach is needed to maximise the opportunities for tourism cooperative organizations and minimise constraints to tourism growth in all the developing countries.

- The tourism cooperative organizations working with other privately managed tourism organizations and its members must ensure full consideration of the issues concerning tourism and related tourism services and must undertake collaboration approach/strategies with other interested government and non-government bodies to promote eco-tourism in developing countries.
- There is imperative need for developing New Model for the cooperative tourism and eco-tourism in developing countries.
- New Generation Cooperative which promote cooperative tourism and eco-tourism can be promoted in developing countries.
- It is essential for developing countries to set up Institute of Eco-tourism for the promotions of the tourism and eco-tourism as in the state of Sikkim and Assam.
- To encourage private investment in tourism projects, the Department of Tourism, in the developing countries must coordinate with different departments for early clearance of the projects and infrastructural development of the tourism facilities.

Conclusion

The Government and Tourism Cooperative Organizations in developing countries must strengthen, promote and create awareness on eco-tourism. It is essential and imperative to sustain tourism products and services within physically and socially fragile environment. For this to succeed it is essential to develop and promote new Generation Cooperative Organizations which take into consideration all the factors of eco-tourism in developing countries.

REFERENCES

Purushottam Jha, (September-2004), "*New Generation Cooperative*", The Cooperator, Vol.42, No.3, pp.101-103.

Ray G.L. (1996). "*Extension Communication and Management*", Naya Prokash Publication, Calcutta, pp. 1-15.

Sami Uddin, (1983), "*Cooperative Sector in India*", S. Chand & Company, New Delhi, pp. 355.

Samwel Kakuko Lopoyetum and Svamani P., (May 2004), "*A Paradigm shift of Cooperative Leadership in the era of Economic Liberalization and Globalization*", The Cooperator, Vol.41, pp. 456–463.

Samwel Kakuko Lopoyetum, (December 2003), "The Dimension of New Generation of Cooperative - 21st Century Perpective", *The Tamilnadu Journal of Cooperative*, Vol.4, No.2,pp.11–14.

Sanjay Kumar Verma, (September-2004), "*Rural Tourism: Role of Cooperative*", The Cooperator, Vol. 42, No.3, pp. 96-100.

24

ROLE OF RURAL WOMEN AS ENVIRONMENTALIST

N. Meenakshi Sundram and *R. Renganayaki*

Introduction

Women constitute half of the population and their development reflects the progress of society. The status of women in society is measured in terms of level of income, employment, education, health as well as the role played by them in family, community and society. They have attained a prominent role in all the spheres viz. agriculture industries, professions like medical, engineering, science and technology, research etc. Still they remain as the most exploited segment of society. Major percentage of women is found in the agricultural sector or in unorganized small industrial sector or in unorganized small industrial units, or as home workers in home based small scale work units. The work involved by the home based workers includes collection of minor forest procedure, beedi rolling, handloom weaving, fishing, leather work and pottery making. Besides women labour in rural India also comprise other type of workers found in Industry, mining, construction and plantation.

Problem of Rural Women Labour

Rural development is conceivable without the active participation of women, who already make up a significant proportion of the workforce and are involved in all economic activities. In spite of the active participation they are marginalized and exploited. An understanding of these exploitative mechanisms is a must to workout the possible strategies towards a transformation in the rural setting.

Table 24.1 : Industries which employed more females than males 1991 (All India)

Sl. No.	*Description*	*Number of Workers*		
		Female	*Male*	*Percentage of Female*
1.	Processing, canning and preserving of fish and similar foods	14215	13938	51
2.	Processing of edible nuts	61761	31092	67
3.	Cotton spinning other than in mills	150159	109718	58
4.	Preparatory operations (including carding and combing) on coir fibres	17975	9471	65
5.	Spining, weaving and finishing of coir textiles	53936	30974	64
6.	Manufacturing of matches	91638	40816	69
7.	Welfare services rendered by organizations operating on a no profit basis for the promotion of welfare of the community such as relief societies, creches, homes for the aged and physically handicapped	170177	99921	63
8.	Domestic services	562396	410527	58

Sources: Registerar General of India, 1991 Census. Women's Link, Vol.6, October-December 2000.

Table 24.2 : Gender composition of the usual status workforce in India

Year	*Rural India*		*Urban India*	
	Male	*Female*	*Male*	*Female*
1961	65.8	24.2	84.8	15.2
1972-73	64.3	35.7	81.3	18.7
1977-78	63.7	36.3	78.9	21.1
1982-83	62.9	37.1	79.2	20.8
1987-88	63.9	36.1	79.0	21.0
1989-90	64.6	35.4	79.8	20.2
1990-91	66.9	33.1	80.0	20.0
1991-92	65.5	34.5	79.5	20.5
1993-94	64.1	35.9	78.8	21.2

Source: Social Welfare Vol.46 Jan. 2000.

Rural Women as Environmentalist

Women working in rural areas knowingly or unknowingly are deeply involved with natural resources, i.e. soil, water, forests and energy. By their own means and ways they develop an under standing about environment. And their own skills are involved in using it for the daily necessities of life. Their tasks in agriculture, the animal husbandary as well as in agriculture, and animal husbandary as well as in the household make them daily managers of the living environment. They have a profound knowledge of the plants, animals and ecological process around them.

Due to unplanned growth, the ecological systems of earth have come to the verge of breakdown. The inherent knowledge of rural women can help avert the catastrophe. A conservation of biodiversity has important bearing in today's context when forces of globalization are altering the whole scenario showing no interest in the sustainable development. In such difficult times, the rural women can play an important role. If she is made aware of her role in the present complex situation. And is encouraged to tackle environmental problems. But this rarely happens. In spite of her closer relationship with nature. She is never considered as an essential part in the process of conservation nor her role is acknowledged.

Strategy for Improvement

The understanding of the processes and reasons of discrimination is, however, the first step towards eradication, them. It is evident that unequal gender relating are responsible for subordinate status of women. In order to achieve development with gender equality. The government should stress on gender sensitive policies and planning.

Steps taken by Government

In order to improve the conditions of women workers the government has taken a number of steps which include enforcement of Minimum Wages Act, setting up of Tribal Co-operative Marketing Development Federation (TRIFED) in 1987.

According to the report of National Commission on Rural Labour (NCRL), self-employment generation programmes like IRDP have made limited impact on the lives of women workers. Women beneficiaries account for only 7 per cent of the total number of beneficiaries and IRDP against the stipulation of 30 per cent. The Government has also undertaken steps to improve the conditions regarding the availability of credit and uni-organisation of women workers. The other aspect of this same story is not that encouraging, the fact being that huge funds for rural poor remain unspent. The State Governments go on crying for more central funds to implement various programmes for the rural poor on the one hand and on the other fail to utilize funds received from the Union Government on this account. This has happened with Indira Awas Yojana. Drought Prone Area Programmes, Desert Development Programme etc. Even the fate for Integrated Rural Development project was no better. Lapses are at various stages the analysis of which could prove useful allocation of funds, restructuring of programmes, efficiency on the part of fund releasing agencies are the various important stages where effective strategies can be worked upon. Also the Gram Sabhas or the village assemblies of the Panchayats can play an important role.

Government of India has set up the National Commission for women with a basic mandate to investigate, examine and safeguard constitutional rights for women. This Commission and the 14 State Women Commissions not only deal with the in-coming cases and come up with recommendations they also work on the lines to generate awareness for rural women. Specially, the women of weaker sections like scheduled castes, tribals etc. Panchayats can effectively help in solving the miseries of rural women workforce by making the welfare schemes (Run by Government) reach them and also by supporting them in their struggle for justice. Women's representation in gram-panchayats is the form of their direct intervention in all related issues.

Need for Multi Pro-Longed Approach

What is needed now is a multi-prolonged approach which focuses itself on effectiveness of state action, promotion of societal awareness and building up of self-confidence and self-esteem among women. The starategy involves state society and self.

There is an increased need for creating societal pressure to treat women as equals and to promote their legitimate interests. In India the qualities that women are required to have to win respect are that they should be passive, partient, sacrificing and modest. But the opposite qualities are needed to assert and have confidence. Women who assert and fight for their rights are harassed, particularly in the rural areas.

The press and voluntary organizations particularly of women should be promoted and activated to secure proper enforcement of law and to bring atrocities against women to the notice of the public at large.

Promoting female education and raising their economic status can be viewed as an effective strategy to give strength and support to women workers. The old stereotype of women as an inferior and exploitable part of family and society must be rejected and changed.

The second prong is sensitization of women of their place in an egalitarian society. Their rights are en-shrined in the Constitution and law. The role of women activists is crucial. They can serve as catalysts of social change especially in the cast of rural women who are excessively oppressed and marginalized due to poor health status improper and insufficient education opportunities, prevalent social evils and labour market discrimination.

A change in the outlook of the men folk and strengthening of the women's movement are the need of the hour. Media is the most effective means of communication and it shall be made to play its role correctly and powerfully on this important and sensitive issue. Committed voluntarty women's organizations should motivate their workers to organize women labour force engaged in agriculture and allied activities in the village by injecting self-respect in them.

Making them literate and acquainting them with the existing welfare schemes of the Goverment and their rights under the Constitution of India. Members of voluntary organizations taking interest in this vital issue should have a strong networking among themselves and their actions should be timely and appropriate to the situation. Such activists or social workers should not pose as superiors but should identify themselves with their less fortunate sisters, suffering under injustice. These "sufferers" should also be mobilised to struggle against discrimination and exploitation.

The weaker employment and wage prospects for women give rise to a vicious circle of low literacy and lesser gainful employment. There is also an urgent need for an alternate strategy to organize the workers in the un-organised sector. Often trade unions and the management question the abilities of women workers to organize, and sad in solidarity with each other. However, the various on-going socio-economic movements organized mainly though women. Stand as a testimony against these allegations. The Chipko movement (Garhwal region), antiliquor agitation (Andhra Pradesh), and Fisher People's struggle (Kerala) in the last few decades are instances of the ability of women to organize. However, many more such ventures are to provide the unorganised women workers the strength and confidence. Assured employment, good and healthy working conditions and maintenance of the minimum labour standards would encourage the women to come forward and participate in the struggle for justice.

Conclusion

In the context of the rural areas, particularly, we have to unite with greater determination. All progressive elements from the society-like judges, advocates, intellectuals and activists, students can play a definite role to create a better world for women who have long been the objects of distress and suffering. Vital of this sensitization their economic status. Poverty and economic dependence are the root causes which lead to poor socio-economic conditions and also the atrocities against them.

REFERENCES

1. Alka Srivastava, "Rural Women", *Social Welfare*, Vol. 46, January 2000 pp.3-5.
2. Chaudhary and Dr. Sangram Singh, "*Department of Extension Education*", "*Rural India*", January 2000, pp. 13-14.
3. M. Srinivasan, "*Economic Census*", "*Women's Link*", October-December 2000, pp. 12-16.
4. Gupta, Registerar General of India, "*1991 Census*", "*Women Link*", Vol.6, October-December 2000, pp. 35–40.

25

SUSTAINED ACCESS TO WATER, CHILD HEALTH AND DEVELOPMENT

S. Sathiyabama and *N. Malathi*

Introduction

Human resource development is the most important approach to raise the levels of aspiration and productivity and to break the vicious cycle of poverty at the most vulnerable points. By releasing this "Creative Spirit" without which no wealth can be created it can rightly be considered as the "Critical input" for realizing the pent up forces of progress. Better health is a part of a balanced socio-economic development in the human environment. Without this all other investment in food production and community services will be less effective. Development recognizes that improved nutrition together with basic services such as health and education, cannot only directly improve the physical well-being and quality of poor but can also indirectly enhance the productivity and their ability to contribute to the national economy.

Children are backbone of a nation. On their health and prosperity depends the health of a Nation. India has the second largest child population in this world next to China. According to 2001 census out of the total population of 104.9 crores, 34 per cent are children upto 14 years of age as against 20 to 25 per cent in the most developed countries. Absence of protected water and sanitation facilities create severe environmental problems.

In order to identify the determinants of child health and the impact of improved water source on child health development a sample of 14 world countries is selected with 4 countries, namely Kenya, Pakistan, Bangladesh and India from low human development counries, 4 countries, namely Srilanka, China, Thailand and Mexico from Medium human development countries and 6 countries, namely Korea, Singapore, United Kingdom, Japan, United States and Australia from high human development countries. In these countries data regarding Human Development Index, Gross Domestic Product (per capita PPP US $ 2001), Physicians per People (per one lakh people), Improved Water Source (%), Infants with low birth weight (%), Infant Mortality Rate (per 1000 live births) and under five mortality rate (per 1000 live births) are collected from Human Development Report (2004).

Determinants of Child Health

Child health depends on Gross Domestic Product Environmental Risk Factors Health Services and Health Status. Health services include health expenditure (preventive and curative care) and Physicians per one lakh people. Environmental risk factors such as supply of improved water and sanitation and Health status in terms of Infants with low Birth weight affect Infant mortality rate and under five mortality rate.

Protecting the environment will be an important part of improving human welfare today and tomorrow. Water is one of the most important environmental factors. Some disease spread through impure water and sanitation. Hence this study focuses on access to water, child health and development.

A study by Rajarathnam (1990) showed that infant mortality is not much related to housing condition (type of house, number of rooms etc.), but highly related to environmental sanitation with respect to drinking water. Water and Sanitation are two non-nutritional factors which have an impact on health status. An attempt is, therefore, made below to look briefly at the water and sanitation situation in the world countries. With a growing population, the pressure to provide water supply and sanitation facilities on a sustained basis is mounting. The basic facilities is availability and accessibility of drinking water and sanitation facilities.

According to Ramesh (1991) the public health condition in

South Arcot District is in noway enviable and dangerous diseases do occassionally visit this district. Illness is highly seasonal and rises dramatically just after the rainy season. Household Environmental conditions are the most conducive for the transmission of malaria, enteric and diarrhea diseases. Among children diarrhea and respiratory diseases are the most frequent causes of death. Plauge is another disease that plauged the district due to insanitary conditions. The stringent measures in sanitation, isolation and treatment of the affected individuals are mass inoculation against the diseases.

It shall be inferred from Table-25.1 that declining infant mortality rate (IMR) from 83 to 3 per 1000 live births is a factor contributing to increasing human development of world countries from 0.488 to 0.946. IMR ranges from 83 to 53 per 1000 live births in low human development countries, whereas these values range

Table 25.1: Indicators of Child Health and Development

S.No. Country	*IMR*	*U_5MR*	*HDI*
Low Human Development			
1. Kenya	78	122	0.488
2. Pakistan	83	107	0.497
3. Bangladesh	51	77	0.509
4. India	67	93	0.595
Medium Human Development			
5. Srilanka	17	19	0.740
6. China	31	39	0.745
7. Thailand	24	28	0.768
8. Mexico	24	29	0.802
High Human Development			
9. Korea	5	5	0.888
10. Singapore	3	4	0.902
11. United Kingdom	5	7	0.936
12. Japan	3	5	0.938
13. United States	7	8	0.939
14. Australia	6	6	0.946

Source: Human Development Report (2004).
IMR — Infant mortality rate (per 1000 live birth) 2002
U_5MR — Under five mortality rate (per 1000 live birth) 2002
HDI — Human Development Index(2002)

from 31 to 17 per1000 live births in Medium human development countries. On the other hand in High human development countries these are very low ranging from 7 to 3.

In case of under five mortality rate (U_5MR), its value decreasing from 122 to 4 is another factor contributing to increasing human development of world countries. Under five mortality rate ranges from 122 to 77 per 1000 live births in low human development countries, whereas these values range from 39 to 19 per 1000 live births in Medium human development counrties. On the otherh and in high human development countries these values are very low ranging from 8 to 4.

Access to Water and Human Resource Development

The link between water, sanitation and health status is a complicated process. These non-nutritional factors have a significant impact on the ability of individuals and households to attain good nutritional status. The World Bank has estimated that more than 30 million life years are wasted annually due to water related diseases. It is estimated that 80 per cent of the diseases and sickness are water-borne.

Water pollution in developing countries creates a major problem of diarrheal disease, not only among children but also among adults. The conditions that facilitate the transmission of these diseases are all related to poverty and unhealthy living conditions. In India, diarrheal diseases are a major cause is less among those using own wells or piped water at home.

Control of these diseases needs to be addressed chiefly through improving sanitation and supply of safe drinking water. Lack of good drainage is another possible cause contributing to water born diseases. Sustained access to water needs the source of improved supply of water and sanitation facilities virtually no municipality has an underground sewerage system. Underground drainage facilities have been provided fully or partially in 16 towns, though the existing drainage systems also suffer from deficiencies. Rural and urban areas have predominanve of open *kutcha* drainage, a rather inefficient medium that is responsible for widespread overflowing, or broken drains.

Table 25.2 : Determinants of Child Health in World Countries

Sl.No.	*Country*	*GDP*	*HE*	*PPP*	*IWS*	*ILB*
Low Human Development						
1.	Kenya	1020	114	14	72	11
2.	Pakistan	1940	85	68	76	19
3.	Bangladesh	1700	58	16	73	30
4.	India	2670	80	51	56	30
Medium Human Development						
5.	Srilanka	3570	122	43	86	22
6.	China	4580	224	164	58	6
7.	Thailand	7010	254	30	90	9
8.	Mexico	8970	544	156	81	9
High Human Development						
9.	Korea	16950	948	180	78	4
10.	Singapore	24040	993	140	100	8
11.	United Kingdom	26150	1989	164	100	8
12.	Japan	26940	2131	202	90	8
13.	United States	35750	4887	279	100	8
14.	Australia	28260	2532	247	100	7

Source : Human Development Report (2004).
GDP—Gross domestic product (per capita ppp US$) 2002
HE—Health expenditure (per capita ppp US$) 2001
PPP—Physicians (per one lakh people) 2003
IWS—Improved water source and sanitation facilities (%) 2002
ILB—Infants with low birth weight (%) 2002

Table 25.2 exhibits that per capita Gross Domestic Product (GDP) has its value increasing from $1020 to $35750 is another factor contributing to increasing human development of world countries. Gross domestic product ranges from $1020 to $2670 per capita in low human development countries whereas these values range from $3570 to $8970 per capita in Medium development countries. On the otherhand in high human development countries these values are very high ranging from $16,950 to $35750.

As the health care facilities (*health expenditure one lakh physicians per people*) increase the IMR and U_5MR decreases in the world countries. Health expenditure per capita (HE) ranges respectively from $ 58 to $ 144 and from $ 122 to $ 544 and from $ 948 to $ 4887 to in low, medium and high human development countries. As the environmental risk factors (improved water and sanitation, infants with low birth weight). Improved water sources (IWS) ranges respectively from 56 per cent to 76 per cent and from 58 per cent to 90 per cent and from 78 per cent to 100 per cent to low, medium and high human development countries. So improved water source increases reduce the infant mortality rate and under five mortality rate.

Infants with low birth weight (ILB) ranges respectively from 11 per cent to 30 per cent and from per cent to 22 per cent and from 4 per cent to 8 per cent to low, medium and high human development countries. So Infants with low birth weight decreases human development is also increases.

Table 25.3 : Matrix of Rank Correlation Between Child Health and Development

Variable	*IWS*	*HE*	*PPP*	*ILB*	*IMR*	*U_5MR*
HDI	0.82	0.95	0.85	-0.67	-0.70	-0.70
GDP	0.82	0.96	0.88	-0.67	-0.82	-0.82
U_5MR	-0.63	-0.75	-0.64	0.66	–	–

Note : Correlation coefficients are significant.
Source : Computed.

The simple method of studying the interrelation between child health and human development is undertaken by computing rank correlation coefficient.

Improved water source is possible with gross domestic product. Gross domestic product increases (ranges from $1020 to $ 35,750), improved water source and sanitation facilities (ranges from 72% to 100%) also increases. It leads to improvement in Human development (ranges from 0.488 to 0.946).

When Health Service (from $ 58 to 4887) increases, Human development (from 0.488 to 0.946) also increases. Rising Gross domestic product (from 1020 to 35,750) leads to higher contribution in the health expenditure ($58 to $4887) and increasing physicians per people (from 72 to 100). It is due to improvement in the human development.

Child health shows inverse relationship between Improved Water Source (IWS) and Under five mortality rate (U_5MR) to extent of -0.63 denoting the result that 0.39 per cent of change in U_5MR is due to the influence of IWS. Existence of inverse relation between IMR and GDP to the extent of -0.82 denotes the result that 67 per cent of change in IMR is due to the influence of GDP. There exists inverse relationship between ILB and GDP to the extent of -0.67 denoting the result that 0.44 per cent of change in ILB is due to the influence of GDP.

Conclusion

This paper enlightens the interrelation between sustained access to water child health and human resource development.

Countries with higher GDP are possible to improve the source of water supply and sanitation facilities (IWS) which has its promotional effect on human resource development.

All the world countries with higher increases the health expenditure, and physicians per people are having favourable effect of child health on human resource development. This is due to reduction in Infant mortality rate (IMR), under five mortality rate (U_5MR) and Infants with low birth weight (ILB).

REFERENCES

Demography, Health and Nutrition; Tamil Nadu Human Development Report (2001).

G.E. Cumber, *Determinants of Health Levels in Developing Countries*, England: John Wiley, 1984.

P.H. Reddy, "Household Toilets and Infant Mortality", *Deccan Herald*, April 11 (1999).

R. Barlow, Health and Economic Development : A Theoretical and Empirical Review. University of Michigan (Unpublished Report). 1977.

T. Rajaratnam and Jyoti, Hallad (2000), "Determinants of Nutritional Status of Young Children in India : Analysis of 1992-93", NFHS Data, *Demography India* 29(2) : pp.179-200.

26

COOPERATIVE ENVIRONMENTAL INFORMATION MANAGEMENT SYSTEM (COOPENVIMS) FOR ENVIRONMENT PROTECTION AND CONSERVATION

M. Karthikeyan and *P. Jagadeesh*

Cooperatives cover all aspects of human life. They are particularly strong in agriculture, consumer, housing, credit workers, transport, and fisheries, and in many other areas in the countries of Asia-Pacific region. Agricultural cooperatives in Japan, sugar and dairy coooperatives of India have created a large variety of services for members besides giving them generous economic returns.

A cooperative enterprise is a form of organization created to meet the needs of the community, in a given situation, without exploitation, as a value-based institution, committed to build a social order characterized by equity, equality and distributive justice. Cooperative enterprises provide the organization means whereby a significant proportion of humanity is able to take its own hands the tasks of creating productive employment, overcoming poverty and achieving social integration. Being a socio-economic institution, cooperatives try to look after both the economic satisfaction and satisfaction of social needs of the members and the community as a whole. More the use of services of the cooperatives, higher is the outflow of social services leading to a stronger social integration in the community.

The qualities such as self-confidence, human dignity, security and a sense of belonging are central to social order, democracy and peace. Cooperatives are the vital component of a civil society with strong ties to local communities. As stated earlier, cooperatives are both economic and social in character—they are business enterprises with a deep sense of social and environment responsibility.

The cooperatives could strengthen our democratic fibre and are the most effective instruments of socio-economic, cultural transformation and environmental protection (Sharma, M.L 1998). "Save Democracy" is the etiquette of cooperatives which also stress about "save nature", "save trees to save man" to protect the environment (*Editorial* TNJC 1991). With their tremendous widerange of activities in the field of production, processing, distribution and all other areas, cooperatives have a significant role to play in the protection and conservation of environment. The reason is that the programme of environmental conservation and protection cannot be expected without the mass support of the people. Cooperatives in India cover almost every village in the country and more than 95 per cent of the rural population. They also exist in almost every important sector of the economy in urban areas. Thus, they have a direct access to a large population and it is easy to muster the public support for the implementation of environmental conservation and protection programme to a grand success.

"Concern for Community" is one of the cooperative principles. It is, therefore, natural to expect that they should encourage proactive stance on the environment protection and take interest in its conservation and protection for the larger interest of the community. However, since cooperatives are user-member driven enterprise, they may take up those issues which are common to both the members as well as the community such as environment, health, youth and women welfare, care for the aged and children, etc. Realising that the cooperatives "existing for the benefit of their members" the ICA in its statement on cooperative identity has emphasized that the cooperatives "have a special responsibility to ensure that the development of their communities—economically, socially and culturally—is sustained". Broadening the scope of cooperatives' work the statement adds that they "have a responsibility to work steadily for the environment protection of those communities".

It is noteworthy to state that cooperatives in various countries are engaged in the protection of environment directly or indirectly.

The forest cooperatives in Canada, forestry and wood cooperatives in the Americas and other European countries, and Kobe earthquake resistance services/disaster management cooperative in Japan are appealing in the activities related to the environmental issues. As for India, sugar cooperatives in Maharashtra took the lead inproviding various community and environmmental services in their area of operation. They have established schools, colleges, hospitals, gymnasia, etc., benefiting not only their members but also the public at large. They have constructed roads for making transportation easier for the farmers (Sharma M.L. 1996). The Bajpur and Nadehi sugar cooperatives have also contributed towards the same social developmental activities and they have installed equipment to check environmental pollution (Veera Nabar 2000). Hospitals and medical colleges are the remarkable contributions made in Kerala alongside the action plan given for the community health through cooperatives by Smita Nair (1994). In Tamil Nadu housing cooperatives are running colleges, polytechnics, ITIs, and rendering environmental services for better being of the community. Gradually other cooperatives like IFFCO entered the field and are rendering laudable service to the environmental development programmes of their communities. IFFCO has covered over 1150 villages where it is providing service like family welfare, adult education, child care, medical and veterinary to check ups, tree planting, sanitation campaigns, etc. Further, to contribute to environment protecton, IFFCO has affording more than 4000 hectares of wasteland and social forestry (Sharma, M.L 1997) and in Tamil Nadu it is advocated to start forestry cooperatives and social forestry schemes in order to control pollution and environment (Ramanujam, K.N 2000). KRIBHCO has diversified into bio-fertilizer since 1995 and also in the area of seed multiplication programme. Bio-diversity conservation through ground level cooperatives and Bio-diversity Cooperative Commonwealth is propounded for joint pricing as well as sharing of the bio-diversity wealth (Khushalani, A.V and Kavita, K. 1999).

The cooperative land development banks are doing wonderful environmental services by means of advancing loans for minor irrigation projects (sinking of wells and energizing water-lifting devices) and major irrigation projects (soil reclamation, land levelling, bunding, field channels, etc.). These banks can render a lot of constructive services to the water management (Chandra Mohan 1999). The banks have also made a beginning in financing forestry as part of afforestation and development of wastelands. Financing to farmers for utilization of scarce ground water resources and management has been done by the Mulukanoor Cooperative in Andhra Pradesh (Veera Nabar 2000). The increasing awareness on environment and rehabilitation of large dams, of Project Affected Persons (PAPs) has led to a rediscovery of a useful base and purpose for cooperatives–organizing PAPs into cooperatives to demand, manage and exploit the reservoirs for fishing, tourism and entertainment (Sunil and Smita 1996) as also the community dugwells in Orissa (Mallik 1996), a lift based canal irrigation system in Andhra Pradesh (Shah 1996), and fisheries with eco-friendly technology for pollution control and income generation through the Mudiali Fisherman Cooperative Society in Kolkota (Datta 1996). Evidences are there about cities for all by housing cooperatives in developing countries to curtail social and environmental problems so as to have safe environment (Khurana, M.L. 1999). This shows that cooperatives have helped in the overall development of rural and urban areas, alongside have helped ameliorate the welfare of the community and protect the environment.

Cooperative Environmental Information Management System (COOPENVIMS)

Cooperatives may adopt the policy of three R's viz., Reduce, Reuse and Recycle to encourage environment upkeep and protection through sustainable consumption pattern. Their objective should be to encourage production and consumption patterns, which would reduce environmental stress. They need to educate individual consumers and organizational consumers to make purchase decisions of environment friendly. Although cooperative movement has a large scope to contribute efforts towards protection of environment, the vast size of the cooperative movement in India cannot be taken too seriously because it is rather a weak movement barring sectors like milk, sugar and fertilizer, which are its strong segments. The absence of proactive management in most cooperatives has been responsible for their callous attitude towards this important concern of extreme social significance. This would need an information system and advertising campaigns, descriptive labelling beamed at various consumer segments called Cooperative Environmental Information Management System (COOPENVIMS).

The Cooperative Environmental Information Management System (COOPENVIMS), which integrates descriptive information (metadata) and data, is a system to capture, store, manage, and distribute information about environmental resources collected, developed, and used by cooperatives and its state and regional partners. COOPENVIMS may be established as a part of cooperative movement and a separate wing of National Cooperative Union of India to have integrated approach to the management of metadata about scientific resources produced or used by the cooperatives.

COOPENVIMS many be developed in response to the increasingly complex nature of environmental assessments. Today's assessments require the integration of human and ecological data and utilize data collected at multiple spatial and temporal scales. The assessments typically involve multiple information management systems and numerous researchers located in geographically dispersed work sites. Successfully completing environmental assessments requires the development of shared inventories of information that support finding and evaluating the information resources for secondary use.

Environment Management Capacity Building Technical Assistant Project as in ENVIS of the Government of India may be launched for the purpose of providing with information on various aspects of the environment besides giving details about the state of environment, the rich biodiversity of the nation and other environmental activities undertaken in our country through a web based information system.

Objectives of COOPENVIMS

The objectives of COOPENVIMS are :

- To create web based databases on the state of environment and its conditions of all regions.
- To establish Cooperative Environment Queries Clearing House to answer environmental queries.
- To establish linkages with environmentalist, researchers, experts and public, and other related national and international organisations at large through knowledge [Centres for Environment Protection (KCEP)].
- To identify environmental information/data gaps and design action plan needed to fill these gaps to have stress-free environment.
- To make arrangements for dissemination of Environmental Information to be reashed to the public.
- To present 'Environment Report' periodically to provide a benchmark for future environmental reporting and also serve as the database for policy making and preparation of environment plans.
- Formulation of suitable management measures through Environmental Management Pan (EMP) for amelioration of adverse impacts.

Green Cooperatives

In order to achieve the objectives of the COOPENVIMS, Green Cooperatives may be organized specially for offering/rendering the environmental services to the community. The Green Cooperatives may render services in the following environmental conservation and protection activities/areas :

Agricultural— Animal Production–Crop Production–Forestry

Building — Private Office Building - Public Office

Building - Warehouse - Historic/Monument - Museum
Defense Site —Air Force Base - Coast Guard Station - National Guard Base - Defense Site
Development—Subdivison - Trailer Park - Apartments/ Townhouses - Office Development - Industrial Development - Medical Facility - Commercial Development - Educational Facilities
Industrial—Degreasing Facility - Coal Gas Facility - Chemical Plant - Power Plant - Wood Treatment Facility - Other Industrial Plant
Miscellaneous— Construction - Church/Synagogue/ Temple/Mosque - Undeveloped Property - Unclassified - Other
Park/Golf—State Park - Private Park - Private Open Space - Public Open Space - County Park - Local Park - Nature Preserve - Golf Course
Recycling—Tire Pile - Salvage Yard - Other Recycling Residential - Single Family
Residential—Multiple Family Residential Retail - Dry Cleaner - Garage/Auto Body Shop - Gasoline Station - Other
Transportation - Railroad - Port/Marina - Airport - Highway
Waste — Dredge Spoil Disposal - Debris Disposal Area - Landfill - Municipal/Industrial - Contaminated Marsh - Spill Site - Contaminated Backfill - Dump-Unauathorized - Drum Storage/Disposal Area - Waste Transfer Station
Waste Water—Sewage Treatment Plant - Spray Irrigation Treatment - Combined Sewer Overflows
Water Production—Agricultural Water Production - Drinking Water Production
Wildlife Area—National Wildlife Area - Artificial Reefs - State Wildlife Area - Private Wildlife Area.

Environmental protection programmes aimed to create environmental awareness and a change in the attitudes in the society with the main objective of making the people to understand environmental awareness and education bring them into direct contact with the environmental problems facing the society that they live and make them think of solutions. The role of cooperative movement in this field of environment protection can be as follows (Neb. D.P 1997):

- Cooperatives are engaged in production, processing, fertilizers and so on, and it is their sense of duty not to harm or disturb the environment;
- Through the network of cooperative education and training facilities they should sensitize their members, directors, leaders, employees and the public about the environment related issues with the hope for transformation of man and society through cooperative way of doing things (Jha, D. and Jha, S.K. 1998);
- Environment protection fund may be allocated from the net profit of cooperatives so as to enable them to advocate in favour of environmental protection and conservation; and
- They may frame and adopt appropriate policies for encouraging eco-friendly business and services.

In order to trim cooperatives and make them fit for fighting the challenges calls for imperatives in the fields of improving productivity through proper utilization of resources such as raw material, manpower, capital, equipment, etc., to bring about reduction in wastes (rejections in inventories) and costs (overheads) so as to have competitive price and quality (Chellappan, K. 1998). To create environmental awareness among the school and college students, "Cooperative Eco-Clubs" may be formed with the objectives (1) to educate the children about their immediate environment and causes for its deterioration; (2) to impart knowledge on the concepts of eco-systems through field visits and demonstrations; (3) to mobilise students towards scientific enquiry into environmental problems; (4) to involve them in efforts to preserve environment.

In addition to above all, the Ministry of Environment and Forests, while refining the National Policy, may consider cooperatives to play a major role in the protection and conservation of environment and due weightage may be given to implement environmental awareness programmes and other related activities as cooperatives are being the prime and effective instruments of the government.

To sum up, with the mass support the cooperatives should engage themselves in the environmental protection and conservation process by establishing a COOPENVIMS is the need of the hour. COOPENVIMS is a set up responsible primarily for imparting knowledge and educating the members, leaders and officials of various cooperatives throughout the nation on the need and importance to conserve and save environment. Similarly, employees also need to be sensitized to environmental issues and their role with reference to their job in the enterprise so that they are able to answer questions on environmental aspects of goods and machinery produced, handled or sold by them. This will enable the entire cooperative effort takers, through COOPENVIMS, to make the people to have a clean and green environment of their own towards cent-per-cent eco-friendly business and service system.

REFERENCES

Chandra Mohan, R., 1999, "Cooperatives in Water Management", *Tamil Nadu Journal of Cooperation,* TNJU: Chennai: 7-9.

Chellappan, K., 1998, "Towards image building by Cooperatives in the changing environment", *Khadi Gramodyog*, 44(7), KVIC:Mumbai: 254-257.

Datta, K., 1996, "Strategies for the fishery cooperatives of tomorrow", in *Rediscovering Cooperation* Vol.11(Ed) R.Rajagopalan, IRMA: Gujarat: 359-368.

Datta, K., 1996, "The Mudiali Society: A model for freshwater fishery cooperatives", in *Rediscovering Cooperation,* Vol. II (Ed) R. Rajagopalan, IRMA: Gujarat: 413-421.

Editorial, 1991, "Towards greener india: Save the Blue Mountains", *Tamilnadu Journal of Cooperation*, 83(1), TNCU: Chennai: 3-4.

Glimpses of Cooperatives, "Recent achievements in Cooperative Sector: Some highlights", *The Cooperator*, 37(2), NCUI: New Delhi: 60.

Govt. of Tamilnadu, 2000, "Announcements of Food Department", *Tamilnadu Journal of Cooperation*, 92(3), TNJU: Chennai: 14.

Human Development Report (10th), 1990, "Globalisation with a human face", in *The Cooperator*, 1999, NCUI: New Delhi: 41-43.

Jha, D. and Jha, S.K., 1998, "The challlenge of leadership for Cooperative Development in the forthcoming years", *The Cooperator,* 35(11), NCUI: New Delhi: 487.

Khurana, M.L., 1999, "Cities for All", *The Cooperator,* 37(5), NCUI: New Delhi: 199-204.

Khushalani, A.V. and Kavita, K., 1999, "Intellectual Property/Royalty rights determination for empowering Bio-diversity conserving Cooperatives and Strategic G-77 and Indian Ocean RIM countries commonwealth", *The Cooperator,* 34(1), NCUI: New Delhi: 35-42.

Mallik, R.M., 1996, "A Micro-Analytical investigation in to community *Vs.* private dugwells: findings from a study of two villages in Orissa", in *Rediscovering Cooperation,* Vol. II (Ed) R. Rajagopalan, IRMA: Gujarat: 25-39.

Neb, D.P., 1997, "Environment Protection and Cooperatives", *The Cooperator*, 35(6), NCUI: New Delhi: 317.

Ramanujam, K.N., 2000, "Concern for Community - New Principal of Cooperation", *Tamilnadu Journal of Cooperation*, 92(3), TNCU: Chennai: 20.

Shah, 1996, "Organisation for the Bhagiratha lift canal project for irrigation and domestic supply: outline of a design and its framework", in *Rediscovering Cooperation,* Vol. II (Ed) R. Rajagopalan, IRMA: Gujarat: 3-21.

Sharma, M.L., 1996, "Community is the main Concern" (Editorial), *The Cooperator*, 34(5), NCUI: New Delhi: 139.

Sharma, M.L., 1997, "IFFCO - One more feather in its Cap" (Editorial), *The Cooperator*, 34(12), NCUI: New Delhi: 327.

Smita Nair, 1994, "Community Health through Cooperatives - An Agenda", *Cooperative Perspective*, 29(1), VOMNICOM: Pune: 42-44.

Veera Nabar, 2000, "Cooperatives on the Upsurge - some success stories", *Tamilnadu Journal of Cooperation*, 92(3), TNCU: Chennai: 29-36.

Vishwanathan, B.S., 1998, "Challenges before Cooperative Sector", *The Cooperator*, 35(10), NCUI: New Delhi: 459.

Websites:

http://www.environment.nic.in/ENVIS.htm#soe
http://www.envis.nic.in
http://envfro.nic.in
http://www.google.com.
http://www.ica.coop
http://www.worldenvironment.com.
http://www.wisc.edu/uwcc
http://www.epa.gov/eims/papers.html

27

IMPACT OF TOURIST ARRIVALS AND BIO-DIVERSITY CONSERVATION OF KODAIKANAL HILLS

V. Dhulasi Birundha and *M. Rajadurai*

Introduction

Tourism has become a major and an integral part of economic, social and physical development. It comprises complete system of nature, the universe, the space and the galaxy, which includes the man and his activities, wildlife, mountains and valleys, rivers and waters, forests and trees, social and cultural system, flora and fauna, the sea.

Now tourism mainly may be means of economy, employment generation etc. for a country. Tourist's point of view tourism makes dreams of sight-seeing, experience, know about the historical accounts, means of hobby etc. However, the flow of tourists to different places of country consists of important components of eco-systems. Those components may be like hill stations, seacoasts, rivers, lakes etc. The tourism connected with these eco-systems often viewed with understanding of that environment and its conservation and also keeping eco-health (sanitation of the environment). This is termed as Eco-Tourism. The flow of tourist population and other anthropogenic activities may cause certain *conflicts with those eco-systems. Here comes the important component of the eco-system, i.e. BIO-DIVERSITY or BIOLOGICAL-RESOURCES.*

The concept of bio-diversity includes the entire biological hierarchy from molecules to eco-systems, or the entire taxonomic hierarchy from alleles to kingdoms, all the populations, species, and all of the different members of all those classes. It also includes the diversity of living interactions and processes at all these levels of organizations. This is such a wide-ranging description that it has kept the definition of the term "bio-diversity" vague and ensured that its measurement remains capricious. Both problems are widely acknowledged among conservation biologist through precious little is done about them. The upshot is that it is still difficult to apply to term the practical purposes such as influencing public policy, or integrating the protection of bio-diversity in to land use planning and management strategies (O.P. Kandari Ashish Chandra, 2004).

Economic Impact on Tourism

Most countries rightfully view tourism as part of their development strategy and as an economic alternative to traditional economic sectors such as agriculture and industry. Rapid development in the means of transport and communication has made the world in to practically one single neighbourhood. Indeed, there are several benefits to tourism. First and foremost, tourism plays a sizeable role in national prosperity. The foreign exchange profits earned by the industry increases reserves and positively affects the balance of payments role of countries. Indeed, tourism it is an important sector in the Indian economy, "where it has become the third largest source of foreign exchange, after readymade garments and gems and jewellery. The foreign exchange earnings from tourism during 1997-98 were estimated to be about Rs. 1.1 trillion (US$ 3,173 million). The rate of growth in foreign exchange earnings from tourism is also high". The Ministry of Tourism of India estimates that the foreign exchange earned by tourism was Rs.142,380,000,000 during 2000-2001.

Employment Generator

Tourism generates opportunities in many sectors, particularly in remote and backward areas. Because it is very labour intensive, it offers employment opportunities to skilled and unskilled workers alike. In other words, tourism has a tremendous employment potential, both direct (travel agents, transport operators, hotels, guides) and indirect (handicrafts, increased demands for foods, clothes, etc.). According to the WTO, total world employment in tourism is estimated at about 20 million. In the case of India as a whole, the Ministry of Tourism estimates that "Rs.1,000,000 invested in tourism created 47.5 jobs against 44.7 in agriculture and 12.6 in manufacturing. Today, the direct employment generated in tourism is estimated to be 16,450,000. Indirect employment is higher at 22,370,000. A sample survey by the World Tourism Council, which uses a methodology called Tourism Satellite Account, concludes that jobs generated by tourism could rise from today's 2.4 per cent to 6.8 per cent in ten years., which means 8,000,000 new jobs (State Government of Tamil Nadu Tourism Policy note, p. 4).

This provides a lucrative field for enquiry in to the details of tourism and environment at Kodaikanal hill-station came in to being in British territories for the convenience of the British residents. It is grown from a small settlement of few houses to a status of a tourists centre and summer resorts.

Ecological Impact

Ecological impacts of eco-tourism are as important as economic but they are not always positive, instead they have more negative impacts. The unregulated and unmanaged growth of floating population. Tourist have to endure heavy disturbance for instance, in protected areas the lions and cheetah decrease their hunting activities when surrounded by six or more vehicles. The expansion of tourism mostly levels behind a trail of litter and destroys the fragile forest habitat and wild life. seasonal influx increases the population, and desities resulted in physical congestion in areas like—beaches, skislopes and historic centres (J.K. Sharma, 2000).

Kodaikanal located amidst the folds of the Verdant Pali hills is one of the most popular serene hill stations in India, which mesmerises any visitor. With her wooded slopes, mighty rocks, enhancing waterfalls and a beautiful lake, Kodaikanal is a charming hill station of South India. Kodaikanal is situated at an altitude of about 2,133m high and covers an area of 21.45sqkm. The hill town is renowned for its educational institutions of international repute. The pride of Kodaikanal is the 'Kurinji-flower', which blossoms once in 12 years. The hill-plantain fruits and plums are known for their freshness and taste.

Table 27.1 : Tourist Arrivals in Kodaikanal

Year	*Domestic Tourist Arrivals*	*% Growth*	*Total Tourist Arrivals*	*% Growth*	*Foreign Tourist Arrivals*	*% Growth*
1993	1427679	–	37110	–	1464789	–
1994	1587274	11.18	40201	8.33	1627475	11.11
1995	1621352	2.15	44675	11.13	1666027	2.37
1996	1625187	0.24	45087	0.92	1670274	0.25
1997	1629462	0.26	45399	0.69	1674861	0.27
1998	1490895	-8.50	40724	-10.30	1531619	-8.55
1999	1532985	2.82	43357	6.47	1576342	2.92
2000	1522802	-0.66	45694	5.39	1568496	-0.50
2001	1547184	1.60	45697	0.01	1592881	1.55
2002	1662616	7.46	50906	11.40	1713522	7.57
2003	1842995	10.85	56774	11.53	1899769	10.87

Source : Tamil Nadu Tourism Development Corporation, Chennai.

Important Places in Kodaikanal

Berijam Lake

Berijam Lake is one of the beautiful picnic spots in Kodaikanal, located 21km away from the bus stand. This lake supplies drinking water to Periakalam town. The scenery around here is delightful.

Bryant Park

Bryant Park situated on the eastern side of the lake, is noted for flowers, hybrids and rafts. Varieties of flowers can be seen in a glasshouse. The annual horticultural show is held here in May.

Shenbaganur Museum

Shenbaganur Museum is located about 5km from the lake. The Sacred Heart College maintains this museum. One of the best orchidoriums in the country with more than 300 species of orchids and wide-ranging collection of birds are found here.

Boat Club

In 1910, a new boat house was constructed, but only from May 1929, few tourists were permitted to avail boat ride as temporary members. Later on they were extended to the public and tourists in 1932. Boats like T.T.D.C., Kodai boats, Carlton boats are available for hire.

Kodaikanal Lake

Kodaikanal Lake is a magnificent man-made lake, which spreads out in a star shape over sixty acres is a main attraction as it is set among the greenery. The boat club here offers several pleasure and racing trips.

Some Recommendations to save Kodai Lake

- Form a Kodaikanal lake management board, and involve all the stakeholders in the Watershed-House Owner, commercial establishments.
- The commercial establishments must immediately pipe their sewage after treatment out of the watershed (schools, hotels, clubs).
- Provide well-maintained public amenities for the public to use and pipe the sewage out of the watershed.
- The toilets around the lake should be connected to a common waste collection system that will pipe the sewage out of the watershed.

Pillar Rocks

Within the city limits one can see three boulders standing shoulder-to - shoulder vertically measuring to a height of about 122 metres providing a panoramic view.

Coaker's Walk

Coaker's Walk derives the name from Lt. Coaker's, who prepared the map of Kodai and it is about a kilometre from the lake, which runs along a steep slope on the southern side of the Kodai. Some of the best views of the plains can be seen from here.

Telescope Houses

To enable the visitors to a panoramic view of the valley and the nearby towns, two telescope houses have been put up at Kodai.

Kurinji Andavar Temple

The presiding deity of the temple is Lord Murugan and is about 3.2km from the lake. In Tamil literature "Kurinji" means "Hilly Region" and "Andavar" means God. Form this temple, one can get a view of the northern plains and Palani Hills.

Silver Cascade

Silver Cascade waterfall is located on the Ghat Road about 8km from Kodaikanal. The overflow of Kodai Lake comes down here as 180 feet high falls.

Bear Shola Falls

An ideal picnic spot located 1.6km from the lake, which can be by a picturesque, rugged hill path. Bears once used to drink water above the falls.

The Summer Festival in Kodaikanal

There are several beautiful hill stations in Tamil Nadu. With the summer Festivals, the hills are even hospitable to welcome the visitors. The summer festival is held in the 'Queen of Hill Stations', the evergreen Ooty; the exquisite Kodaikanal or the salubrious heights of Yercaud. Cultural programmes, adventure sports, boat races, flower and fruit shows add to the splendour of the Festival.

Why Bio-Diversity Conservation

When human needs interacts and reacts amongst themselves and their surrounding non-living environment an ecosystem created.

The value of bio-diverse livelihood and the genetic improvement of locally important crops will be the prime importance for conservation of bio-diversity.

For determining priorities for bio-diversity conservation/ preservation six criteria are suggested by Flint (1992) :

- Value for developing countries,
- Diversity and distinctiveness,
- Threat,
- Gaps in other programme,
- Comparative advantange livelihood of success.

Possible Solution for the Sustainable Tourism

For sustainable utilization of tourism potential the growth of this activity must be there without causing irreversible damage to the natural environment, activities relating to tourism development should take the following into consideration:

- Carrying capacity of this tourist must be assessed for tourism promotion
- Development of tourism should be in harmony with environment
- Tourist must be given with sufficient awareness about the fragility of the ecology of that locality to be visited by them

- Tourist must be involved in conservation efforts
- Public policy must be reformulated
- Nature sites require protection and maintenance
- Mass-media can do service on this concept.

Importance of Bio-Diversity Conservation

- Bio-diversity can become a much-needed link to provide continuum in their biological world.
- It is the backbone of agriculture, animal husbandry, forestry selection and adequete utilizations.
- It helps in redesigning of landscape and reconstruction of ecosystem.
- Best tool to establish a much-needed bridge within the social and cultural world.

Conclusion

In order to protect and conserve the natural environment of Kodaikanal, it is the prime responsibility of the Goverment to arrest deforestation activities. Steps must be taken by the Government to regulate and control the rapid construction activities. Environmental awareness campaign should be launched periodically. These measures will go a long way in enhancing and protecting the natural environment of Kodaikanal.

REFERENCES

Department of Tourism Government of Tamil Nadu, Tamil Nadu Tourism Complex, Chennai-2.

J.K. Sharma, 2000, *Tourism and Development: Design for Ecological Sustainability*, Kanishka Publishers, Distributors, New Delhi.

O.P. Kandari and Ashish Chandra, 2004, *Tourism Biodiversity and Sustainable Development*, published in 2004 in India by Isha Books, D-43, Prithiviraj Road, Adarsh Nagar, Delhi-33.

Prabhas Chandra, 2003, *Global Eco-tourism-codes Protocols & Charters*, Kaniskha Publishers, Distributors, New Delhi.

V. Dhulasi Birundha, 2003, *Environmental Challenges Towards Tourism*, Kanishka Publishers, Distributors, New Delhi.

28

ENVIRONMENTAL ISSUES IN CORPORATE BUSINESS

M. Sivasubramanian and *I. Sundar*

Key Words

Sustainable development, certification, code of conduct, corporate social and enveronmental responsibility, eco-efficiency, greenwash, stakeholder accountability, triple bottom line, voluntary initiatives, win-win strategy.

Introduction

Globalization of capital flows has heightened the pressure to improve corporate governance and responsibilities around the world. Many business enterprises have made huge investment in production and distribution of goods and service that have resulted directly in extensive waste and degradation of natural resources or encouraged consumption patterns that do the same harm. It could be observed that logging and mining enterprises, pulp and paper mills, agribusiness, oil, chemical, cement, iron and steel companies as well as many other enterprises and imdustries have degraded large areas of tropical forests, marine and coastal resources, fresh water resources, agricultural land and the urban environment, and affected global climate and the ozone layer. This is due to the narrow focus on objectives such as market share and profitability, much of big business, has disregarded environmental as social aspects which are central to the concept of sustainable development.[1] Given their size and global reach, transnational corporations (TNCS) have often been singled out as major culprits of unsustainable development. Today this sector comprises approximately 60,000 parent firms with over half a million affiliates which in 1997 accounted for one-third of world exports (UNCTAD 1998; 1999). Many companies, business associations, governments and international organizations are adopting a discourse and policy suggest that corporations can simultaneously make profits and be good citizens. There appears to be a growing recognition that the increasing freedom enjoyed by big business in the era of globalization needs to be complemented by increasing responsibility (UNRISD, 1995).

In some countries prosesses and policies associated with globalization and neo-liberal reforms have weakened the regulatory capacity of certain state institutions. Similarly, at the international level, various attempts to influence transnational corporations practices through codes of conduct have been abandoned. In a context when such institutions have weakened, corporate self-regulation and voluntary initiatives have evolved dominant approaches to promote business responsibility.[2] It is significant to note that increasingly big business during 1990s embraced the language of corporate social responsibility and took measures to reform management system and make them more responsive to the environmental and social concerns of different stakeholders, This paper examines the nature of changes that are taking place in corporate discourse, policy and practice related to business responsibility and its implications for a sustainable development.

Approach to Corporate Responsibility

The contemporary wave of corporate responsibility is different in various respects. Its effects can be felt in many countries, the forces driving it are more varied, the dominant policy approaches and institutions arrangements for promoting business responsibility are different and until quite recently environmental aspects have been more visible than social ones.

Today certification is one of the business responsibilities. It is an evaluation system intended to provide proof of a company or a product's environmental and social performance. Hence some questions arise concerning the degree of independence of the verifier and the rigour of the standards and indicators used to measure performance. There is a bias with respect to labeling of eco-friendly products between developed and developing nations. It should be noted here that major criticism of some forms of environmental certification relates to the fact that what is evaluated is environmental management not environmental impact[3] does your company have an environmental policy, not, to what extent has your company reduced its emissions of use of energy. The relationship between improved environmental management and impact is not always as apparent as one might think.[4]

The code of conduct is a business responsibility in relation to sustainable development. It is a set of ethical principles and standards that attempt to guide a firm's environmental and social performance. The formulation of code of conduct by companies and industry or business associations has escalated sharply during 1990s. This suggests that business may be crafting a new relationship to the environment and society.

It is significant to note that business initiatives in the field of corporate responsibility are particularly evident in relation to drawing up code of conduct or specific policy statements on environmental and social aspects. In recent years an increasing number of companies and business and industry associations have developed code of conduct and guiding principles. Despite the current wave of global enthusiasm for codes, the proportion of companies adopting them is still relatively low in most countries. Employees and consumers are often unaware of the existence of company codes and firms frequently fail to specify the nature of sanctions for non-compliance of a particular concern. Effectively, self-assessment or independent verification of compliance with codes is rarely practised by companies (Dammen 1999, ILO 1999 and UNCTAD 1996).

The usefulness of the international code should be assessed not only in terms of the number of countries which have implemented it, but also in terms of its being a tool to raise and maintain public awareness of the issue and to help build public pressure on corporations to change their inappropriate marketing practices. Corporate social responsibility is an important aspect of sustainable development of business and human society. It is often used in a broad sense. Historically progress associated with corporate social and environmental responsibility has been driven to a large extent by state regulation, trade union activism and collective bargaining. An increasing reliance on voluntary initiatives may be undermining these driverse corporate responsibility, such as initiatives are often presented as effective alternative to state regulation, while their success in many industrialized countries is often involved in an imporant regulatory component (Porter and der Linde 1995).

By focusing on the specific initiatives taken and the innovations introduced by individual companies in the field of corporate environmental and social responsibility, a considerable attention can be made on investment, production and marketing trends and their environmental and social implications. It is important to retain a sense of perspective of the broader trends related to corporate policy and performance. Many companies focus narrowly on one particular aspect associated with corporate responsibility, for example environmental protection, and ignore others such as labour conditions and indigenous rights. In relation to environmental certification some international trade unions concerned with such instruments are legitimizing the activities of companies that continue to abuse certain basic rights.

Eco-efficiency is another aspect of corporate responsibility. It is a process of adding more value while steadily decreasing resource use, waste and pollution (Schmidheiny and Zorraquin 1996:5). Eco-efficiency is the dominant model of environmental management reform promoted by the World Business Council for Sustainable Development. The model has been criticized as promoting technical fixes that do not fundamentally alter an economic system that continues to generate environmental degradation and social exclusion. It is as much an approach to promote economic growth and competitive advantage as it tends to encourage environmental responsibility. The leadership role is very essential in the field of corporate environmental responsibility more specifically in the eco-efficiency model actively promoted by business associations like the World Business Council for Sustainable Development. It has to influence and shape international and national environmental policymaking and such activities have been diversified through promotion of corporate social responsibility. The leadership role is also evident in the way industry influences international bodies. Global corporations are increasingly influenced by policymaking at the international level and related to environmental and development issues (Krut and Gleckman 1998 and Dawkins 1995).

It is crucially important for environmental improvements to be measured against externally defined benchmarks of sustainable environmental governance in setting up frameworks for corporate environmental activity and assessment and monitoring to ensure that industry moves towards global sustainable development (UNCTAD 1996: 86-87).

The corporate responsibility is diluted by greenwash activity. It is a process of disinformation disseminated by an organization so as to present an environmentally responsible public image. In other word it is an attempt by corporations to hide the unpleasant environmental facts about their activities by adopting an environmental discourse or specific policies and practices that appear to be environment-friendly, but have done nothing to change the relationship of business to the environment (Greer and Bruno 1996). It is difficult to assess the significance of contemporary trends associated with corporate environmental and social responsibility by simply weighing up selected cases of best practice or greenwash. Today an increasing number of companies are not only adopting innovations associated with corporate environmental and social responsibility, but are also shifting from a narrow piecemeal and technical approach to one that is holistic, systemic and more responsive to multiple stakeholders. Some corporations are socially and environmentally destructive and attempt to preserve and expand their markets by posing as friends of the environment and leaders in the struggle to eradicate poverty (Corporate Watch 1996). Accordingly some corporate policies and initiative wanting substance are not implemented but considered with a system that encourages business as usual (Welford 1997). The role of international organization for standardization in upsetting national standards is essential in dealing with corporate responsibilities. It specifies environmental management system that may be objectively audited for self-declaration of third party certification.

Here concerns arise regarding the fact that what is measured is not whether a company is actually reducing its negative environmental impact but whether it is utilizing an environmental management system. There is also concern that the ISO process for developing environmental standard can be heavily influenced by business interests and there may be little participation from developing countries and NGOs.

Stakeholder accountability is another responsibilty of corporate business toward sustainable development. They are affected by the achievement of the organizations objectives (Freeman 1984: 46). Corporate stakeholders include employees, neighbours, public interest groups, customers, suppliers, governments and the general public (Schmidheiny 1992, 10). Stakeholder responsibility refers to the need for companies to be responsive to the needs and concerns of their different stakeholders.

In reality only a few some stakeholders are taken into account. This reflects not only the practical limits of corporate responsiveness and variations in the legitimate claims of differencing stakeholders on a company, but also their relative power to exert influence and pressure. The analysis of the structural underpinnings of corporate environmentalism and stakeholders accountability suggest that it is likely to be a very uneven phenomenon in both sectoral and geographical term. Developments associated with flexible specialization, global commodity chains and foreign direct investment affect some product sectors and countries more than others. Furthermore, each commodity chain can assure very different characteristics in terms of the actors and market conditions that shape the possibility of improvements in environmental managements.

The protection of brand name image and company reputation has become key managerial concern in certain product sectors (Nelson 1996). To minimize any tarnishing of reputation, some companies not only attempt to respond to environmental stakeholders concern related to business practices that have put them in the spotlight, but also anticipate where the result problem might come from and take preventive action.[5]

Triple bottom line is one of the aspects of corporate responsibility. Here companies should be concerned not only with the traditional bottom line associated with profitability, but also goals related to environmental protection and social needs. This notion suggests that companies are seriously attempting to address the negative environmental and social effects of their operations and move beyond compliance with government regulations. There is a concern, however, that companies are using their concept in situations where the goal of profit maximization outweighs other company objectives.

Regarding corporate responsibility, voluntary initiative is one of the aspects. These encompass a widerange of initiatives that go beyond existing laws and legislations related to environmental and social protection. They may be unilaterally developed by industry designed and run by government, jointly developed by government and industry (UNEP 1998). An increasing number of companies are engaged in voluntary initiatives. Such initiatives, however, lack of substance and credibility they are not effectively implemented, because they are not subject to independent verification and are often promoted as alternatives to what may be more powerful drivers of corporate responsibility such as government or international regulations.

It should be noted that voluntary initiatives are often drawn up without regard to principles of transparency, independent verification and worker participation as well as core international standard to protect labour the environmental and human rights. Usually voluntary initiatives are mainly adopted in richer industrialized countries. Their apparent success in a number of countries has been contingent on certain institutional conditions associated.

It is often assumed that voluntary initiatives should and can be adopted across very different product sectors with globalization the increasing global reach of transnational corporations and changing forms of corporate organization structures, extensive and complex product or commodity chains have developed in many sectors (Gereffi, *et al.,* 1994). Participating in these chains are very different type of actors, with very different capacities to respond to environmental and social concerns and subject to very different pressures and social concerns and subject to very dirrerent pressure, opportunities and constructs (Volts Moltke, *et al.,* 1998). The feasibility of adoption of voluntary initiatives across the board and their capacity to translate into improved corporate environmental and social performance can a considerably in different sectoral contexts.

Finally, win-win strategy is a major responsibility of corporate business firm. It is a corporate strategy that enables a company to simultaneously improve in environmental and social record while reducing costs and increasing competitiveness and productivity. The possibility of win-win strategy suggests that companies should adopt relevant management reforms as part of a rational business response (Murphy and Bendell 1999). Further there is less need for the so called command and control regulation. The win-win notion has been criticized for marginalizing the role of governmental regulation and civil society pressure as key drivers of charge, and for exaggerating the proposition that environmental management reforms is necessarily good for profits.

It is often argued that corporate environmental and social responsibility is basically a rational business to ecological constraints and market opportunities. There is a considerable debate regarding the win-win supposition that environmental improvements can go hand in hand with cost reduction. For certain firms particularly small and medium sized enterprises, this may not be the case. Even if the cost-reduction component of win-win scenario is weak, there is still the possibility of enhancing competitiveness. Indeed this is probably one of the main strategic advantages that companies can gain from improvements in environmental and social policy and practive. But it is not clear whether this means the promotion of sustainable development. To the extent that improved competitiveness can be gained largely through imagery as opposed to significant improvements in a firm's environmental and social performance.

Conclusion and Policy Measures

It could be seen clearly from the above discussion that in the absence of governmental and international regulation and more concerted, coordinated civil social pressure, the process of promoting corporate environmental and social responsibility in developing countries will remain lukewarm at best. In order to promote environmental and social responsibility among corporate business firms, the following suggestions can be considered :

1. Adequate and timely information about company products and processes needs to be made available to public to allow effective tracking and assessment.
2. Good corporate governance should be developed in encouraging socially responsible trade development.
3. Efforts should be made to prevent the disinformation disseminated by an organization so as to present an environmentally responsible public image.
4. Precautionary principle should be undertaken in relation to environmental challenges and there is an urgent need to encourage the development and diffusion of environmentally friendly technologies.
5. These should be measures to overcome the inertia of the present system of production and to create a new momentum towards sustainable development.
6. Initiatives associated with corporate environmentalism should encourage consumers to adopt very different consumption pattern that would significantly reduce environmental degradation.
7. Efforts should be taken to encourage markets for environmental goods and services such as eco-labelled products,

organically produced foods, biological food products, recycling—friendly packaging, eco-tourism, environmental auditing and certification services and cleaner technology.

8. Massive flow of foreign direct investment should facilitate transfer of cleaner technology.
9. Companies need to incorporate social and environmental values into their policies and operations. These values need to be integrated into companies' definitions and measurement for progress and success.
10. Public participation is one of the most important requirements for promoting corporate business responsibility towards sustainable development. Here NGOs should stress the active participation of stakeholders especially those in the communities directly impacted by a company's operations.

NOTES

1. To define sustainable development, the Brundtland Commission referred to development that meets the needs of the present without compromising the ability of future generations to meet their own needs (WCED, 1987). This, of course, is a tall order, which suggests that sustainable development should be viewed more as a processes than a goal. Another formulation might be a process that minimizes the contradictions between economic growth, environmental protection and social well-being (Utting, 1993).
2. The terms business responsibility and corporate responsibility are used in this paper as convenient shorthand for corporate environmental and social responsibility, although there are, of course, other aspects of responsibility related to business activities.
3. This criticism has been levelled, in particular, at ISO-14001 (see Krut and Gleckman, 1998), and has emerged more gererally in some of the literature on corporate social responsibility (see Hopkins, 1997).
4. As a possible explanation, Levy (1997:60) suggests the fact that larger firms have more power to resist the introduction of costly environmental investments or bureaucratic inertial.
5. This point was stressed by several participants (notably those currently or previously connected with large oil companies) at a workshop organized by UNCTAD (20 May 1999) on corporate social responsibility.

REFERENCES

Colchester, M. "The international tropical timber organization: Kill or cure for the rainforests?" *The Ecologist*, Vol. 20, No. 5, 1990, pp. 166-173.

Dawkins, K. Ecolabelling: Consumer's right to know or restrictive business practice? Mimeo, Institute for agriculture and trade policy, Minneapolis, 1995.

Dommen, E. Pertinence et limites des codes ethiques, mimeo, Geneva, 1999.

Freeman, R. *Strategic Management: A stakekholder approach*, Pitman, Boston, 1984.

Gereffi, G.M. Korzeniewics and R. Korezeniewics, "Introduction: global commodity chains," in G. Gereffi and M. Korezeniewics (eds.,), *Commodity Chains and Global Capitalism*, Greenwood Press, London, 1994.

Greer, J. and K. Bruno, *Greenwash: The Reality Behind Corporate Environmentalism*, Third World Network, Penang, 1996.

ILO, Voluntary Initiatives Affective Training and Education on Safety, Health and Environment in the Chemical Industries, Sectoral Activities Programme, ILO, Geneva, 1999.

Krut, R. ISO 140001: A missed opportunity for sustainable global industrial development, *Earthscan*, London, 1998.

Murphy, D., and J. Bendell, Partners in time? Business, NGOs and sustainable development, Discussion paper no. 109, UNRISD, Geneva, 1999.

Nelson, J. Business as partners for development: Creating wealth for countries, companies and communities, The Prince of Wales Business Forum, London, 1996.

Porter, M. and C. Van der Linde, "Green and competitive: ending the stalemate," *Harvard Business Review*, September-October, 1995, pp. 120-134.

Schmidheiny, S. Changing course: A global business perspective on development and the environment, MIT press, Cambridge, Massachusetts, 1992.

Schmidheiny, S. *Financing change: The financial community, eco-efficiency, and sustainable development*, MIT Press, Cambridge, Massachusetts, 1996.

UNCTAD, Self-regulation of environmental management: An analysis of guidelines set by World Industry Associations for their member firms, UNCTAD. Geneva, 1996.

UNCTAD, World Investment Report, 1998, Trends and Determinants, UNCTAD, Geneva, 1998.

UNCTAD, World Investment Report, 1999: Foreign Direct Invesment and the Challenge of Development, UNCTAD, Geneva, 1999.

UNEP, Voluntary Initiatives for Responsible Enterpreneurship: A Question and Answer Guide, Industry and Environment, Vol. 21, Nos. 1-2, Jan.-June 1998, pp. 4-9.

Von Molkte, K. *et al.*, Global Product Chains: Northern Consumers, Southern Producers and Sustainability, Environment and Trade 15, UNEP, Geneva, 1998.

Welford, R. Hijacking Environmentalism: Corporate Responses to Sustainable Development, *Earthscan*, London, 1997.

INDEX